William Henry Allen

Stock keeping for amateurs

A manual on the varieties, breeding, and management of pigs, sheep, horses, cows

William Henry Allen

Stock keeping for amateurs
A manual on the varieties, breeding, and management of pigs, sheep, horses, cows

ISBN/EAN: 9783337198244

Printed in Europe, USA, Canada, Australia, Japan

Cover: Foto ©berggeist007 / pixelio.de

More available books at **www.hansebooks.com**

Stock Keeping for Amateurs:

A MANUAL ON

THE VARIETIES, BREEDING, AND MANAGEMENT OF
PIGS, SHEEP, HORSES, COWS, OXEN, ASSES,
MULES, AND GOATS,

AND

THE TREATMENT OF THEIR DISEASES.

DESIGNED FOR THE USE OF YOUNG FARMERS AND
AMATEURS.

By W. H. ABLETT

Author of "Farming for Pleasure and Profit," "Arboriculture for Amateurs," "English Trees and Tree Planting."

LONDON :
"THE BAZAAR" OFFICE, 170, STRAND, W.C.

LONDON:
PRINTED BY ALFRED BRADLEY, 170, STRAND, W.C.

Stock Keeping for Amateurs.

CHAPTER I.

PIGS.

Introductory Recommendations—Varieties of Pigs—The Chinese Pig—The Suffolk and Norfolk—The Shropshire—The Rudgewick—The Cheshire—The Old Sussex—The Hampshire—The Berkshire—The Tonquin—The Essex Half-blacks and the Essex and Hertford—The Dishley—The Old Irish—Neapolitan, Maltese, &c.—Advice in Buying—Considerations in Pig Keeping—Accommodation for Pigs, the Pig Stye—Labour in Superintendence—Feeding—Breeding—Improvement of Breeds—Descriptive Terms.

INTRODUCTORY RECOMMENDATIONS.

PIGS, as a rule, are not animals that are profitably kept by farmers, when the amount and value of the food they consume is taken into account. They, of course, under any circumstances, eat much that would be otherwise worthless of itself, and could be put to no useful purpose, which in time gets converted into so much good pork or bacon; while they make a large amount of valuable manure in the process—results that are apparent to everybody. The difficult matter, however, is to make pigs pay when there is no large amount of coarse, or half-spoiled food to resort to. This, however, can be done by those who give sufficient attention to the subject.

There is no art in their management, when the farmer

B

thrashes out a whole stack of "tail wheat," which is expressly devoted to these animals' delectation, the art consists in finding these hearty eaters enough to enable them to subsist and leave a profit to their owner.   To do this successfully, much depends upon the breed of pig selected.   Some are chosen for the large size they attain, and different breeds are held in estimation by different people, for some point of excellence or other, which possesses value in their eyes, with a definite object in view.   The main qualifications, however, to be regarded in the choice of a breed with a view to their paying are that they be hardy, and not sensitive to changes in the weather, that they will eat coarse rough food, and fatten quickly, when it becomes desirable to fat them, and these qualifications are to be found in the improved Berkshire.   They are very hardy and will eat almost anything; changes of weather affect them less than almost any other breed, and when they are nearly ready for the butcher it takes comparatively very little to put them in first-rate condition for killing. It is true the best bred ones do not attain a large size, but it is one most suitable for porkers, and they fetch long prices from the pork butcher; and this ready means of disposing of them gives the opportunity of clearing off a good deal of the stock at times, when the supply of inexpensive food, which in ordinary seasons can be scraped together, happens to fall short from some cause or other; but with good management this seldom needs to be the case.

Experience has shown that store pigs, when kept simply as store pigs, seldom pay between the ages of two months and twelve months old.   Of course, the price of pigs varies at times, like everything else; but in average years a pig at two months old generally fetches a pound, and when a number of breeding sows are kept, and the pigs are sold off at that age, and at that price, a very handsome profit can be secured.   Between two months and twelve months, when the pig is growing (we are speaking now when it is fed only upon food that costs very little), if the value of its food is estimated at a shilling per week (the particulars of which we will shortly allude to in detail) for the forty-two, or forty-four weeks remaining to make up the year, the nominal value of the pig would be supposed to be three pounds two or four shillings, including the first pound, its estimated value at two months; but store pigs would not be worth nearly so much as that in the condition they would then be found, from living upon hard fare; and therefore it would be a loss to keep them, unless their manure was very desirable, or there was a quantity of rough food for them that was wanted to be eaten up; standing, therefore, upon the simple merits of feeding, it is clearly the most paying plan to dispose of them at the age of two months, just after the time they have left the sow, and have become thoroughly weaned and accustomed to the separation, when they will have attained a certain value in the eyes of a purchaser as being fit for young stores.

VARIETIES OF PIGS.

There is a very large number of different breeds of pigs, and it would perhaps be impossible to name them all, but some species stand boldly out, being remarkable for conspicuous features, which we will briefly refer to.

The Chinese Pig.—Perhaps we are more indebted to Chinese pigs than to any other variety for the improvement effected upon the original European stock of swine. There are several kinds comprised in this breed, but the most distinct are the white and the black, the white being better shaped than the black. The best are very white skinned, the hair being thin, and also white, with short head and small ears, very thick neck, and high chine, the head in a fat pig appearing buried in the fore part of the body. The legs are short, and the belly nearly touches the ground. The leading characteristic of the breed is to accumulate fat, which causes them to be bad bacon pigs, the flesh being very tender, so that a great improvement is effected by crossing the pure Chinese pig with an English breed, which results in the addition of more lean flesh.

The black, which are similar to the white in their main points, are more prolific, and are quicker growers, and attain greater weights than the white Chinese. It is, however, when they are young, and eaten as porkers, that their meat is found to be

B 2

unrivalled in flavour; and the breed is therefore held in high estimation by dairymen and others who have a quantity of skim milk to use up, which, mixed with barley meal, is the best food that can be given to them, and, fattening quickly, they do not take so large a quantity as many other kinds of pigs would consume. When they attain a large size their flesh is coarse, oily, and unpleasant, and their principal value is in their making the best roasters.

There is also a small kind known as the "small white breed," which never attain a large size.

The Suffolk and Norfolk.—These are of two distinct kinds, but they so closely resemble each other that they are invariably spoken of as the same breed. They do not attain a large size, but are held in good estimation in some districts, as being hardy and prolific; but they are useful only as porkers, and do not turn out good bacon hogs.

The Shropshire.—This is a very large breed, and is held in high estimation where a stock of bacon forming the principal article of food in the shape of meat consisting of bacon is wanted for consumption on a farm, for the great size it attains, as it turns out enormous flitches. The change in the manner of living, however, when farm servants are now not commonly fed at their employers' table, has caused this description to be less freely resorted to of late years, and it is rapidly going out of fashion.

The Rudgewick is another large sized animal, which is met with chiefly on the further borders of Surrey and in Sussex, but they do not appear to be commonly known throughout the country. Their large size as bacon hogs causes them to be chiefly valuable on this account, and they do not demand any particular care and attention in rearing.

The Cheshire.—The original Cheshire pig attains to a very large size, and is by no means a good looking animal, having a big head, with large flapping ears, round and narrow back, and long bony legs. Bacon made from this variety used to be commonly met with in Manchester and the neighbouring large factory towns, where it was largely consumed by the working population, but the breed is now comparatively seldom met with. In old times people wanted to see how large they could cause their pigs to grow, and sacrificed quality to quantity, very often this description having been known in individual instances to attain the weight of 12cwt.

The Old Sussex is a well shaped pig, which does not attain a very large size. They grow freely and come to maturity early, but possess no special points beyond those we have indicated.

The Hampshire enjoys a better reputation, perhaps, than it really deserves, being a somewhat coarse, raw-boned animal, and the favour in which it is held possibly may be traced to the methods used in curing it, and to the mast which abounds in the New Forest and other woods, where these pigs pick up a large amount of food, just as they did in the days of the Saxon Heptarchy. The old stock, however, is being fast obliterated having been crossed extensively in late years by other varieties, in which the Berkshire has played the most prominent part.

The Berkshire we have spoken of in laudatory terms in our introductory remarks, and although it is not commonly regarded as a breed which attains a large size, yet it produces animals of very great weight occasionally, which alteration in size must be attributed to the gradual improvements which have been effected by judicious crossing, to which the Chinese and Tonquin breeds have mainly contributed.

The Tonquin, to which we have alluded above, often called "Tonkey" or "Tunky," a corruption of the word "Tonquin," is said to have originated from a cross of small Berkshire and Chinese pigs. They are small in size and well shaped, and as pork they are not excelled by any other description of pig. They are generally white in colour, thick, and compact in shape.

The Essex Half-blacks, and the Essex and Hertford.—Both bear a very high reputation as good kinds of pigs, which are distinguished for certain good points that are prized by many.

The Dishleys are handsome pigs, and are the type of perfection in the shape sought to be attained by pig breeders, springing originally, it is said, also from crosses between Chinese and Berkshire ; they have several defects which render them objectionable, as they are very tender and subject to be unfavourably affected by changes of weather; they also take a great quantity of food to fatten properly; the sows are bad nurses, and they are not very prolific; when fat, however, they are masses of flesh, and appear to be "scarcely able to see out of their eyes for fat," as is often commonly expressed; being nearly the same height, length, and thickness, their general contour being lost in a mass of fat.

We will close our brief notice of the varieties of pigs by mentioning the Irish and Mediterranean breeds.

The Old Irish—In its original form the Irish breed of pigs is a large and coarse one, ugly in shape, narrow in body, with large ears, and very large bones. The breed, however, has been vastly improved of late by judicious crossing with some of the best English varieties, and Irish bacon and pork, which used some years ago to be considered very inferior, now take rank with the average qualities of English bacon. Great attention has been paid to the improvement of stock in some parts of Ireland, where the preparation of bacon for the English markets is now a very large trade.

Neapolitan, Maltese, &c.—In the Mediterranean countries there are some very useful breeds of pigs to be met with, which resemble one another very closely in their general characteristics, the principal of which are the Neapolitan and Maltese.

Many of our English breeds are indebted to these for improvement; as regards the smaller kinds, particularly the black varieties, their delicacy of flesh and beauty of form having been considerably improved by judicious crossing with the Neapolitan.

They are symmetrical in shape, round and plump, small in bone, fine snout, with longer head but shorter body than the Chinese pig, which they resemble in many points; but they are rather larger in body, almost destitute of hair or bristle, and coal black in colour. Their aptitude to fatten at an early age exceeds that of almost any other kind of pig, while the flavour of their meat is very superior. They are fairly prolific, good sucklers, and may be kept so as to thrive on food of moderate quality only.

ADVICE ON BUYING.

From the foregoing the reader in search of information will be able to gather the leading characteristics of the various kinds of pigs mostly reared in this country, and he will have to take into consideration, when making his first purchase, his own position and circumstances, in relation to pig keeping—whether he will have to sell them as fat roasters or porkers, as bacon, or to breed for sale as store pigs.

A dairyman who does not rear calves, and has a large quantity of skimmed milk to dispose of, would find the small breeds, which fatten quickly, answer his purpose the best. But if the farmer resides in the neighbourhood of the mining or manufacturing districts, the larger breeds will, in all probability, be in much greater request, whilst the London market will be the best for small kinds of pork, and these considerations must direct the first choice of breeding stock, together with the class of food that is likely to be produced, and general facilities of feeding.

Success, or non-success, depends much more upon such considerations than they generally obtain credit for, and many people, instead of duly weighing them, take the first breed of pigs that comes to hand, perhaps because somebody else has done well with them, whose position and circumstances, relative to the points to which we have referred, are entirely different to their own, or perhaps for no reason at all, but that they want to keep pigs, and think one sort will suit as well as another.

In purchasing, a young sow is preferable to an old one, less money is generally asked for her, and the purchaser will have the prime of her life. She should have a capacious belly and not too great an inclination to fatten, free from natural defects, and with at least twelve teats.

Many sows become too fat during their period of gestation, but this may be overcome by restricting their food somewhat. An old sow is invariably coarse about the teats, and wears an old look, and they are sometimes, from the rough usage they have experienced during the course of their lives, at times bad tempered. In the case of a young sow, with kind treatment, she may be made very quiet by the time of farrowing, which will be a great point in her favour. The sow should be chosen good at the shoulders and loins, and long in body, with short legs.

In the case of the hog, the chief points insisted on by good judges are breadth of chest, width of loin, chine, and ribs, depth of carcase, and compactness of form, accompanied with docility and general beauty of appearance. Head not too long, the forehead being narrow and convex, with fine snout, full cheeks, and small mouth; the ears short and thin, with sharp pendulous ends pointing forward, the *tout ensemble* of the head denoting good temper and sprightliness.

The neck should be full and broad, especially on the top, where it should join broad shoulders and broader chine; the ribs, loin, and rump being of uniform breadth, ending in a tail not too low nor too long, so as to be unseen at top when the animal is fat. The back of a well-shaped hog is straight, or slightly curved, the chest is deep, broad, and prominent, the ribs well set and springing well from the chine, the shoulders widely extended, the thighs inside and out should be very thick, and the belly, when pretty fat, should nearly touch the ground, which entails the possession of short legs. The bone must be fine and joints small, the hair long, thin, and fine, having few bristles, and the skin thin and supple, without looseness. The colour should always be uniform.

The breed of the hog is of more importance than that of the sow, as the progeny take after the male rather than the female, and there may be reasons at times for the latter not being too highly bred.

CONSIDERATIONS IN PIG KEEPING.

Some excellent points in the choice of pigs for breeding were given in the *Field* newspaper a few years ago by Mr. Jno. Coleman, Professor of Agriculture, Albert Veterinary College, as well as good hints upon their management. The letter referred to was in answer to the queries of a correspondent who sought information upon the following points:

1st. What breed is best to keep?

2nd. What might he expect to make by each sow per year, with luck and care?

"I will do my best to enlighten him upon these points," says the Professor, "premising that my views are based upon practical experience in pig breeding.

"In deciding upon the description of animal, the pig breeder must be influenced partly by the wants and prejudices of the locality for which he breeds. There is in many places an antipathy to black pigs, for instance, from a notion that the meat is not so white and delicate, and, however suitable a black breed might prove in other respects, it will not do to keep such, unless we are bold enough to attempt conversion—a very slow process. Again, in countries where the black breed prevails white animals are not popular, being considered less hardy. We must, therefore, breed for our market. Further, our choice of large, medium, or small breeds must be guided by the same rule. If the pigs are wanted for pork, then the small breeds, either white or black, are best, coming early to maturity, and laying on fat rapidly. If, on the contrary, large bacon is required, as is the case when a coarse feeding population has to be supplied, the large breeds will be most profitable. And if there is a market both for bacon and pork, then the medium-sized animals are best.

"When black pigs are in favour I would recommend the Berkshire breed, for the following reasons. They are hardy, and will pick up their living during the greater part of the year; are fairly prolific, and produce animals suitable for either pork or bacon, the latter being noted for the large proportion of lean flesh. Other breeds may fatten as rapidly, but none cut up better, and few so well. There is no advantage in having high-bred sows; such animals are generally inferior breeders, and especially is this the case if much forced when young. Select average specimens of the sort with good breeding points; such are, length of body, straight back, rather fine head, and tapering neck, depth fore and aft, twelve well-shaped teats, and a fair covering of long soft hair, which indicates quality and constitution. It must be borne in mind that the sows will at times have to live hard and rough it, getting their own living in the grass or stubble, and only be cared for as the time of farrowing approaches, and whilst suckling; therefore it is of great importance to select a hardy breed, and the Berkshire are in this respect unequalled.

"We now come to the second question, viz., the value of the produce. This will depend upon two points—the prolific nature of the sow, and the market price of the stores. Suppose we take four litters from each animal, commencing with the young sow when a year old; we cannot estimate the average of each litter above eight—the first is often only four or five, and is no criterion of the breeding powers, as I have known cases when two or three the first time, were succeeded regularly by ten or twelve. Of the eight pigs farrowed, six on an average will be raised, and by good management and some luck, two litters may be produced annually. Perhaps we can hardly count on this regularly, but we may assume it, having estimated our produce moderately. The value of the offspring at eight weeks old will vary from 15s. to 20s. a head. It follows, then, that we may expect a gross return per sow of from £9 to £12 per annum. At three years old the mother, unless she prove an extraordinary breeder, may be replaced by a hilt, and the value of each will be about the same.

"What is the cost of keep? This will vary according to the resource of the farm and the management. On an arable farm, or one consisting of arable and grass, there will always be a certain proportion of damaged produce, decayed roots, &c., which but for the pigs would be wasted, and upon such food a hardy sow will, with the addition of a little wash, or merely water, keep herself in good enough condition from the time of weaning until within three weeks of farrowing again. Sometimes it may be good policy to add a handful of beans to the wash; but if allowed range of ground this is often quite unnecessary. Therefore, I think, we may fairly assume that during half the year the sow supports herself, and repays us in her manner for what she consumes. During the remainder of the year, however, it is a very different story. We must improve her condition before farrowing by generous diet, which shall encourage the secretion of milk, and when she has farrowed, we must force on the young litter through her by every means in our power. Giving a some-. what rough calculation, inasmuch as no exact experiments have as yet been made on this point, I should consider that at least 3s. to 4s. per week for twenty-six weeks would be required to do justice to the litter, and this does not include attendance, which, with straw, may fairly be put against the manure. In round numbers, the expense of maintenance will range from £4 to £5, whilst the value of produce may be taken at from £9 to £12, so that we may look for a clear profit per sow of about £5 to £7. If this can be done, I venture to predict that your correspondent's sows will prove the most paying sort of his live stock, and that this is frequently realised, and often exceeded, I am convinced.

"One point of importance remains to be noticed. We must contrive that our litters shall drop at periods when the weather is not extreme, thus : Midsummer and Christmas are bad times ; the beginning or middle of July, and about February will be the best seasons, both as regards weather and requirements of the market. The autumn litter will be fit for winter pork, whilst the spring litter will run on, and come in for bacon the following season. It will be impossible to always regulate the farrowing as above, and, indeed, it may not be always desirable ; but it is well to avoid as much as we can the dead of winter, or height of summer, as the pig is, of all our domestic animals, most sensitive to atmospheric influences."

ACCOMMODATION FOR PIGS.

Having selected the breed of pig to be kept, the next question that must be considered carefully is how to house him to the best advantage. As to rear and feed pigs properly a good deal of trouble has to be incurred, the sties should be constructed upon such a principle as to enable the pigs to be fed quickly and to be kept clean. The pigsty should always face the south. One man or boy could look after a great number of pigs, if the animals were housed in a way to save extra work. The sties should be placed upon a slight inclination, so as to allow the urine to drain off, and keep the animals dry, which is a great point in pig economy. The sties should be fifteen feet long, and seven wide, part to consist of a boarded house, or brick or stone erection (according to the materials at command), which, while snug and warm, yet ought to be well ventilated. A fair sized opening should be made for egress and ingress of the pig, but this in cold weather it will be found desirable to contract, by fitting the opening with sliding boards to run in a groove. The uncovered part should have but a low wall or railing only, of sufficient height to keep in the pig. In this open portion the trough is usually placed, where the pigs of one litter are generally fed indiscriminately, that is, they are allowed to feed themselves so; but even in one litter it will be often found that the pigs vary in size and strength very much, so that generally one or two little ones get shouldered aside by the heavier ones, and pushed away, when, with a deprecatory squeak, they will run to try their chance at the opposite end of the trough, only to meet the same rebuff, perhaps, from a strong pig which has placed itself there. The anxiety and worry these little pigs endure has a very unfavourable effect upon their growth. Where there is only a paling, and not a thick wall, all this can be avoided, as well as the great waste of food prevented, which sometimes takes place through the pigs putting their feet in the trough—slipping into it and by pushing it over in their haste to get at it—by placing the trough, which should have a flap at the top to turn over at pleasure, *outside* the sty, in the sides of which are holes only large enough to admit the head of one pig at a time. Not only waste will be avoided (for often men can scarcely help pouring the wash, or barleymeal, or whatever the food may happen to be, over the heads of greedy animals), but the troughs can be cleaned out more effectually, and all knocking of the pigs about, which is sometimes done by an angry man, is avoided by the adoption of this plan. In most old-fashioned pigsties a trough is placed inside the wall of the open portion, and the man who feeds the animals pours out of a bucket the contents into the trough. The result of this procedure is, as we have before stated, that the pigs, in their eagerness to get the food, crowd round, and half of it is poured over their heads, or on the ground, and wasted. This may be obviated by having a spout carried through the wall, into which the food can be poured, and so save the inevitable loss which occurs through the slovenly plan we have mentioned, if the previous one we have suggested cannot be adopted. A boiling and baking house should be placed at the high end of the sty, and a cesspool to receive the drainage at the other.

Some breeders have adopted for fattening hogs a form of sty which is but little larger than the size of the pig. In these no litter is allowed, the chewing of which is thought to be injurious to the pig; the floors consisting of boards, placed upon an inclined plane, having holes in them for the water to drain off, and these are swept out every day.

Both in the covered and uncovered divisions of the sty brick or stone floors should be used, both sloping towards the drain, the inside being raised a few inches higher than the outer division.

Although pigs will eat the dirtiest food with the greatest possible apparent relish, and appear to enjoy wallowing in the worst filth they can meet with, it is a vulgar error to suppose they thrive best when dirty. An occasional washing or brushing will keep their skins in good order, and cause them to thrive better than when neglected. The latter is easily performed, and docile pigs enjoy the operation immensely.

Labour in Superintending.

In properly constructed sties a good deal of labour and trouble is to be saved, so that one good strong boy or man could attend to a great number of pigs; but no stock pays better for being well looked after, and great saving is to be effected by *constant* attention. As we shall describe hereafter, pigs can be fed economically on a good deal of waste stuff, often regarded as rubbish. This is often thrown to them all at once, when sufficient pains *are* taken to get it together, and they waste more than they consume; when, if a portion of it only were tossed over to them occasionally, the extra labour incurred by the additional attention

would be more than compensated for in the saving effected, and
their feeding being made more constant and regular, the pigs
are contentedly employed during the whole day, and a cheerful
placid temperament is encouraged in the animals.

FEEDING.

The getting together of food, either for sows or young store
pigs of the hardy breed we have indicated, can be managed
in a variety of ways that are often entirely overlooked. It is
bad policy to starve any animal, but a very small quantity of
good sound food is to be supplemented by a large amount of
refuse which the majority of farmers never are in the habit of
regarding as food of any value. When turnips are singled, and
mangolds thinned, and weeds hoed, it is generally customary to
allow them to lie on the land where they have been cut out.
These should all be collected together, and carted off to the
pigsties; they should not be thrown into them all at once, when
the pigs would eat a portion and trample down the remainder,
but they should be put down *outside* the sties and tossed over
several times during the course of the day, which anyone can do
when passing the sties from time to time. What they do not eat
their hoofs will convert into manure, which, although of lesser
account than manure made from richer food, yet possesses a
certain value, and is, of course, mixed with other manure which
has been made from more fertilizing matter by pigs which have
been receiving better food. Everything that can be got together
in this way should be carried to them. The mowings of lawns,
the trimmings of banks and hedges, garden refuse, all will be
found useful for this purpose. The pigs will readily eat young
nettles that may be cut from the hedges by a boy with a hook, and
the labour in performing these jobs will be amply repaid by the
result. In some parts, where acorns are plentiful, it pays well to
give a shilling a bushel to the wives and children of the labourers
for gathering them, and a large amount of sound food is got by
this course. Many people say that acorns are bad things to give
to pigs, as it hardens the flesh ; and, although it is not expedient

to give any large quantities of acorns to quite young pigs, they will do no harm to old sows, or strong store pigs; any ill effects which might arise from this item of food would be removed by the final course of fattening, when they are finished off at last with barley meal, mixed with potatoes.

A large copper should always be kept going to boil up messes of green stuff, which can be sweetened and made palatable by a few handfuls of malt dust, pollard, or even bran; sweepings of corn from markets, damaged corn, or any other refuse that can be had for a mere trifle in some places according to situation.

At certain seasons, when no green stuff is obtainable in any quantity, hardy pigs can be kept on split mangold. The benefit arising from giving the mangold merely split instead of being cut up is that the pig has to masticate it, and thus properly mixes it with his saliva, which is essential to the proper assimilation of food.

We have spoken of economical contrivances for feeding pigs, but pigs are commonly fed upon turnips, mangold, and potatoes in winter, and upon grass, clover, and other trifoliated plants in the summer, beside picking up a good deal in the fallow fields under the supervision of a boy to prevent them from straying, or getting into mischief.

As stated before, a copper should always be in use for boiling any green stuff that can be got together during summer, and roots and refuse in winter, which, mixed with a few handfuls of pollard, will be eaten readily by them, and upon which they will thrive, simply as store pigs of course.

Porkers, however, which are designed for the butcher at an early age, require a better description of food, and are generally fed upon skim milk and barley meal, the latter made into a thin paste or even mixed with the milk. Upon this fare there can be no question the pigs thrive remarkably well, and the very best quality of young pork is turned out, but where calves are reared, as we have mentioned elsewhere, the skim milk can be more profitably employed. If the class of pig is one that fattens easily, the cost of the barley meal can be sensibly reduced by mixing it with boiled potatoes, upon which they will thrive, but they must be finished off with the best possible food that can be given.

Damaged rice can frequently be bought cheap, and has been largely used by some people, but such food as tallow chandlers' greaves should never be given to pigs, as it communicates a rank taste to their flesh, which should be avoided. Malt dust, bakers' sweepings, bean dust, the sweeping of corn markets and railway depôts, all come in for pigs when they are obtainable, and do admirably well to mix with green and other food that is grown upon the farm. It should always be borne in mind that pigs thrive better upon cooked food than upon uncooked, and the cooking, if made a matter of *routine*, soon ceases to be looked upon as a troublesome operation.

Like every other animal the pig thrives best when fed with regularity, and this should be done three times a day.

A great many diseased potatoes may sometimes be bought at threepence and sixpence per bushel, according to their badness, and these it will be found more advantageous to *bake* instead of to boil. The operation of baking dries up the moisture, and the pigs will eat them readily and enjoy them as much as they would do corn. A convenient method of baking them is to build a small kiln with a few bricks, and to lay a plate of iron on these. The potatoes are laid on the plate, and the fire can be made from any old rubbish that is laying about, with a few handfuls of coal to light it in the first place ; the ashes will be found very useful in drilling in seeds, if carefully saved and put in a dry sheltered corner.

In feeding pigs most economically all small potatoes, carrots, parsnips, the fallen fruit from the orchard, can with advantage be used, as well as the haulms of potatoes, peas, or any similar vegetable offal. Brewers' grains, if they are obtainable at a cheap rate, are an excellent addition to all these, which, if sedulously gathered together, will be found to go a long way towards feeding the pigs.

In manufacturing businesses the very smallest savings in material and labour are always taken into account, and a variation of $2\frac{1}{2}$ per cent. often makes all the difference between loss and profit on a transaction. When markets are glutted, and large manufacturers are unable to sell their goods at a profit, the only rational step to take would appear, to those unacquainted with the details of a manufacturing business, to cease manufacturing. But the manufacturer knows that to stop making means to lose the hands which he has got together, and that machinery deteriorates in value from standing idle. It is no easy matter, therefore, to make up for all these inconveniences when the trade turns again, and it is found more desirable to go on, and even to sell goods at a small loss rather than stop ; and while there is no necessity for a farmer to lose money by what he rears, yet he may often profitably imitate the example of the manufacturer in attention to trifling details, and effecting a saving whenever he has the opportunity. In handsomely appointed pig sties, where the owners pride themselves upon doing things well, large quantities of clean fresh straw may often be seen thrown down for the pigs. Chopped straw is a valuable article of food, the details relating to which will be found in another place, and litter of a much less costly nature can often easily be got together for pigs, and answer quite as well. In some places large quantities of leaves are obtainable ; nothing forms better manure, and instead of allowing these to blow all over the place and make it untidy, they should be frequently swept up and stored away for litter. When collected in small quantities they dry better, and are less likely to heat than when a good many are got together at once.

The writer upon one occasion had about forty pigs, which were kept upon this economical system, which he showed to an old friend, a Suffolk farmer. The latter laughed incredulously when he was informed that a large portion of the food of the pigs he saw consisted of nettles, boiled and unboiled. "I know," said he, "that pigs will nearly keep themselves at certain seasons upon grass, and I am in the habit of turning mine out occasionally—but nettles ?—I never heard of such a thing!"

To convince him, some cabbages were pulled up from the garden, a little grass mown, and some young nettles cut out of the hedge, and these were all thrown over into the sty together, and the pigs picked out the nettles and eat them up first. And

why should they not ?   Young nettles are often eaten as a vegetable in some of the northern districts of the kingdom, while
"nettle tea" is frequently used as a useful spring medicine by
the labourers.   But a vast amount of prejudice exists against the
use of such things, and there are many waste substances that can
often be profitably utilised in a similar manner, one of the chief
obstacles to the proper use of which is the labourers, who would,
as a rule, much sooner run to the miller, or to any stores at hand
for the pigs' food, rather than trouble themselves with such
economical contrivances; but the attendants should be made to
follow, to the letter, the instructions which are given them.

Pigs which have kept up their growth and gradual improvement upon rough, indifferent food, immediately make rapid
progress when put upon better-class provisions, and are fattened
off much quicker than when they have been accustomed to food
of richer quality, that has been supplied to them in quantities
adapted only for the condition of store pigs.

BREEDING.

Those who are in the habit of buying their pigs, do not get
an uniform quality of stock, which is always desirable, and it will
be found better practice to raise one's own, which can be very
soon done, no animal of those kept on a farm in a domesticated
state being so prolific, the fecundity of the sow, and the early
maturity of her offspring, being unexampled in animal economy.

Both the boar and the sow will be in a fair condition for
breeding at the age of ten months.   If used earlier, the genera-
tive powers of both will become deteriorated, the boar becoming
stunted in his growth, and prematurely wearing signs of old age,
while the sow soon becomes worn and feeble, and produces
unhealthy litters.

The period of gestation lasts about sixteen weeks, or 112 days,
and generally the months of April and October are chosen, as
nearly as possible, for mating, as the sow, if put to the boar
in the latter part of the above named months, will bring her
litter at the end of August and February respectively.   In the

first case the young pigs will get sufficiently strong to be able to go through the winter, and in the latter they will get strong enough for summer grazing, when there is a good deal to be picked up out of doors in the stubble fields, and amongst other waste and litter of the different crops.

Some breeders aim at producing five litters of pigs in two years, but it is too much for the sow to bear, and an average of four litters is quite enough, as a reasonable period of rest is necessary for the well-being of the animal, and, although the sow will take the boar soon after farrowing, it is not expedient to allow her to breed again so quickly.

The sow, during the time of her gestation, is allowed to run about with other pigs, and she seldom slips her young, unless she meets with some injury or other, or has by chance taken something in the way of food which has disagreed with her, which at times will happen when she has eaten too many roots. A sharp blow across the nose, which an ill-tempered attendant may perhaps give, will cause premature farrowing, but in the ordinary way she can run with the others till within a week of the time of her parturition.

She should then be put into a convenient sty, or pen, and fed upon soft food, as wash, whey, swill, &c. ; and as the time approaches she should be carefully watched, and her bed made of short dry straw, for, if a well strawed bed is prepared for her, there is a danger of the little pigs being smothered, or the sow overlooking them and lying down upon them.

After she has littered the sow should be nutritiously fed with skimmed milk, warm wash, or whey, mixed with meal or bran, and the young pigs should be allowed to feed with her, so that they will learn to eat, and shift for themselves by the time they are weaned.

They should be castrated or spayed while they are sucklers, this operation being performed before they are six weeks old, from a month onwards, so that they may be weaned at eight weeks, when they will generally have recovered from the operation.

After they have been taken away from the sow, the young pigs should be fed at least three times a day with good food of

a farinaceous order, mixed with milk, wash, or whey, and given
warm.   In a week, or a little more afterwards, they will begin to
eat potatoes, when the warm food may be gradually discontinued
and cold feedings substituted, until they are strong enough to
rough it with the rest, if the litter comes at a time of year when
pigs are turned out, and at four or five months they become what
are termed store pigs.

Where young pigs are intended to be disposed of fat, it will
aid their fattening if they are occasionally washed, which can be
soon done by an experienced hand.

The litter falling in February, or at any period in the early
spring, will require careful attention during cold weather.

IMPROVEMENT OF BREEDS.

In making a selection of a sow for breeding purposes the
first consideration is usually to obtain an animal of large,
long body, that will weigh well in the scale, whose progeny
is likely to be numerous, and attain a large size.   But from successive breedings with this main object in view, the home stock
may be gradually getting coarse, and under these circumstances
it will be a good plan to cross the sows with a boar of some of
the smaller varieties which possess the good points that are
required.

Boars retain their powers for many years, and where one is
used for a long time, a smaller one should be kept, as large boars
are sometimes very dangerous with small sows.

Boars should be fed off at about three years old—or a little
longer.   After this time they are apt to feed slowly and be
unprofitable for the butcher.   The boar should at all times be
carefully treated, especially when he begins to get older than
three years, for as they advance in age they occasionally turn
savage, especially when they have been thwarted by a sow being
taken from them.   A blow from a boar's tusk has been known
to rip up the side of a sow, and he occasionally becomes a
very dangerous animal to those who are in the habit of looking
after the pigs, and on this account some breeders are in the

habit of sawing the tusks off. A strong cord is put in the mouth and fastened round the upper jaw, care being taken that the cord is placed over the tusks, so as to prevent it slipping, and the tusks are then sawn off.

DESCRIPTIVE TERMS.

The boar is called indiscriminately by that name, or brawn, or hog; when castrated he becomes a gelt, or gilt, cut-pig, hog-pig, or barrow pig. The female a sow, or open-sow, when young, as a distinguishing term, when there is no doubt about her ultimate disposition, and when spayed a gelt or sow-pig. In the other stages of their growth, applied to both male and female, they are termed store-pigs, fatting-pigs, sucking-pigs, and sucklers.

CHAPTER II.

PIGS (CONTINUED).

Diseases of Pigs, and their Treatment—Diarrhœa—Colic, or Spasm of the Bowels—Inflammation of the Bowels—Diseases of the Spleen—Inflammation of the Chest and Lungs—Catarrh or Cold—Protrusion of the Rectum—Diseases of the Urinary Organs—Quinsy or Strangles—Phrenitis, or Inflammation of the Brain—Skin Diseases—Scrofula—Measles—Mange—Rheumatism—Fattening,—Feeding for Shows—Weight and Measurement—Killing—Cutting Up—Scalding and Singeing—Smoking—Ringing.

DISEASES OF PIGS AND THEIR TREATMENT.

THE diseases of pigs are not very numerous, and a good many of these are brought on by carelessness and want of attention ; and it is fortunate that this is the case, for the pig is a very awkward patient to administer any curative treatment to, for in consequence of its screaming, it is almost impossible to drench it, as there is great risk of choking the animal if a draught is poured down its throat while screaming.

Owing to the voracity of the pig, the digestive organs are most commonly the seat of disease ; but as the symptoms of most of these disorders are somewhat obscure, an attack generally has made some little progress before it is discovered, which causes cures in serious cases to be matters of difficulty.

Diarrhœa.—Diarrhœa is not uncommon, and should be checked at an early stage, for if continued for long it sometimes ends in inflammation, and becomes highly dangerous. A good remedy will be found in the following :

Powdered opium	15grs.
Prepared chalk	4drs.
Powdered ginger	1dr.
Peppermint water	4ozs.

These quantities will be sufficient for about eight doses, one dose to be administered twice a day while the disorder continues. The fæces should be examined, and if found to be slimy a dose of salts should be administered.

Colic, or Spasm of the Bowels.—This is not a very common disorder, the symptoms being those of sudden and violent pain, which are evidenced by the manner of the animal. Tincture of opium, and spirit of nitrous ether combined—double the quantity of the latter to one part of the former—should be administered in doses according to the size of the animal, commencing with one drachm of tincture of opium to two of spirit of nitrous ether, in the case of a small pig; up to eight drachms of the former and sixteen of the latter in that of animals of the largest size, given in warm water. If relief is not obtained from this treatment, bleeding is resorted to.

Inflammation of the Bowels.—Inflammation of the bowels is most frequently produced by unwholesome food, and is much more common than colic, there being two phases of the disorder —acute and sub-acute. In acute inflammation there is considerable pain, as in colic, but without intermission, accompanied with a good deal of fever and loss of appetite. In sub-acute inflammation the symptoms are of a milder character.

Bleeding is usually resorted to for this disease, the vein on the inside of the foreleg being opened, and a flow of blood allowed, in proportion to the size of the animal, from two ounces to two pounds.

If the vein in the inside of the foreleg cannot conveniently be opened, the tail may be cut; but as the object is to ensure a copious flow of blood, a few drops merely will not be sufficient.

Purgatives of an oily nature, such as linseed oil, should also be administered to relax the bowels. If constipation prevails, an injection should be used, and in addition a warm bath will be found beneficial, which, in the case of small pigs, can be easily given. In the milder form, from two to five grains each of calomel and opium will be found efficacious.

In constipation of the bowels, where there is not inflammation, linseed oil will be found a useful medicine, which the pig will often take without any great degree of persuasion being necessary, and of its own accord, and to it may be added a few drops of croton oil when the constipation is very obstinate. Epsom salts, Glauber salts, and infusion of senna, are all good medicines to administer ; but, as they need to be given in the form of a drench, are sometimes a little troublesome. Jalap, in doses of from one scruple to a drachm, accompanied with six to twelve grains of scammony, will be found more convenient.

Diseases of the Spleen.—There are several diseases of the spleen which are dangerous, particularly rupture (which, strictly speaking, is not a disease) and inflammation—rupture is fatal, and inflammation very dangerous. The symptoms are obscure, but they are generally indicated by vomiting, coughing, foaming at the mouth, and grinding of the teeth.

Bleeding and purging is the course of treatment resorted to, though the chances of a cure are extremely problematical.

Inflammation of the Chest and Lungs.—This disorder is commonly brought on by neglect, and is often caused by the wet and unsuitable places in which the animal is lodged.

The disease assumes both the form of pleurisy, which is attended with pain, and of bronchitis, with cough and phlegm. Bleeding in both cases is resorted to, though it is more imperatively demanded in the case of pleurisy.

The presence of either disease is indicated by quick breathing, fever, and loss of appetite. As well as bleeding, the bowels should be moderately well opened, but the animal not too much purged, and the following administered once a day :

Calomel	...	...	...	...	...	...	...	...	...	1 to 3grs.
Tartarised antimony		...	...	...	...	...	1 to 3grs.
Nitre ...	...	...	...	...	...	...	...	...	5 to 20grs.

After one or two doses, the calomel may be omitted, and if the disease is well defined, blisters can be applied to the chest.

Catarrh or Cold.—Colds arise in the same way, and are produced by the same causes, as with other animals, the principal symptoms being a cough, and discharge from the nostrils. With good warm housing and care, the animal soon recovers, but, in severe cases it is necessary to give medicine, of which the following will be found a suitable description :

Antimonial powder	2 to 6grs.
Nitre	10 to 30grs.
Digitalis	1 to 2grs.

which should be administered several days in succession.

If the disorder extends to the lungs and becomes bronchitis, the case gets more serious, and the animal needs to be bled. A stimulant should also be rubbed on the brisket.

Protrusion of the Rectum.—Young pigs are rather frequently attacked with this disease, which often ends fatally. Those which are kept in towns, and whose food chiefly consists of strong animal substances, abounding with gelatine, and not so well tempered with vegetable compounds as that which forms the food of country-bred pigs, are most subject to it. It is, however, sometimes produced by violence. The best treatment is to keep the subject clean and quiet, and not allow it to have any other food but a little milk, so that the bowels may be tolerably empty before the gut is returned to its place. The pig being firmly held in position, the parts should be washed, and the rectum returned and pushed up some little distance. Some strong thread doubled several times, should then be passed through the anus and tied with a knot, and the animal kept on milk for some days, no solid food being given to it.

Diseases of the Urinary Organs.—The pig is seldom subject to diseases of the urinary organs. Inflammation of the kidneys is somewhat uncommon, but when it does occur, bleeding is often resorted to, aperient medicines should be given and a warm bath. Inflammation of the bladder requires to be treated in the same manner, while a dose of opium in addition will assist in allaying the irritation.

Quinsy, or Strangles.—Fat hogs are the most subject to this disorder, which, if not relieved, ends very often in suffocation, to which it owes its latter name. The throat swells up, the breathing is rapid, and the pulse accelerated, while the tongue is covered with saliva, and gangrene follows. Immediate treatment is called for, and bleeding and purgative medicines of a cooling

nature resorted to.   The swelling may be reduced by puncturing
it, or by the application of a seton.

The Epidemic.—Pigs are likewise subject to "the epidemic," as
it is termed in other animals, and often proves very troublesome,
the local symptoms being lameness in the feet, arising from
soreness between the toes, and inflammation of the parts which
connect the bone with the horn.   Pus is thus formed, and the
hoof cast, while fever prevails in the system.   Epsom salts should
be administered and an astringent, consisting of a saturated solution of sulphate of copper or zinc, or the preparation which is
commonly used for foot rot in sheep applied to the feet.

Phrenitis, or Inflammation of the Brain.—This disorder, which
occasionally attacks pigs, is indicated by dullness, and sometimes
even by blindness, and at times violent convulsions.   Bleeding
and purgatives are the most suitable remedies.

Skin Diseases.—Pigs are rather subject to diseases of the
skin, which often arise from high feeding.   A cooling lotion may
with advantage be used.   A good application is formed of the
following :

Muriate of ammonia ... — ... — ... — ...	4drs.
Acetic acid ... ... —	1oz.
Cold water	1 pint

Scrofula.—Scrofulous diseases mostly occur with animals that
are very finely bred, or too much bred in-and-in.   There is no
cure for this, and the only way to prevent it is by changing the
boar occasionally.   Tubercles are found in the lungs, and in the
mesentery, which last interferes with the absorption of the chyle,
which ultimately causes the animal to dwindle away and die.

Measles.—Measly pork is sometimes heard of, the disease
known by this designation in swine having its seat beneath the
skin, where are found a number of small watery pustules, of a
reddish colour.   There is also fever, cough, discharge from the
nostrils, and pustules under the tongue.   It generally yields to
cooling medicine, such as Epsom salts and nitre, and attention to
feeding, and is very seldom fatal.

Leprosy is a formidable skin disease, but it is very rarely heard
of in this country.

Mange.—Mange is occasionally met with in pigs, but much

less frequently than with cattle, horses, and sheep. It is, however, sometimes seen, being characterised by itching and the usual symptoms. A local application is the best treatment, such as the use of tobacco water, sulphur ointment, or mercurial ointment, well rubbed into the skin.

Small-pox is extremely rare with pigs.

Rheumatism.—Pigs are very often troubled with rheumatism, which is generally produced by exposure to cold, and by damp sties or bedding, which attention and good housing mostly prevents. From two to five grains of colchicum, repeated daily for three or four days, is the best treatment to resort to.

As may be seen from the foregoing, due attention to the wants and comforts of the animals are the best safeguards against disease in pigs.

FATTENING.

Fattening being often a special transaction by itself, we have not included it under the head of ordinary feeding; many breeders, not fattening their hogs, disposing of their pigs either while young or in store condition.

Fattening is the most important part of the business of pig keeping, as the profit hinges upon the rapid and satisfactory way in which this can be done. The almost universal method of fattening pigs is with barley meal, mixed with water into rather a thin paste, and given three times a day. Nothing can be better, as they are pushed on fast by it, and the meat produced is very fine, but the great cost is against it.

The enormous quantity of food that is sometimes given to hogs when it is designed to make them fat, is almost beyond belief to those who have studied this part of the business under its most economical aspects, and it then becomes no wonder that the question is often raised whether pigs can be made to pay.

It is commonly estimated that porkers will consume, while fattening, from two to three pecks of corn food. If a large full sized hog, he is supposed to consume from one and a half to two, or even two and a half bushels per week; the calculation being that his weight will increase at the rate of 9lb. or 10lb. per bushel—roughly speaking, if of the size of fifty stone ; and one which fats with tolerable readiness, may probably gain in weight at the rate of two stone per week. About six sacks of barley, and one of peas, are supposed to be necessary to fatten a hog of sixty or seventy stone. This is only supposed to be a moderate calculation, for in the case of those hungry breeds of pigs which take a large amount of food to fatten, the consumption will be considerably greater.

Pigs, however, of a kindly sort, which have been brought up from their infancy upon rough fare, will fatten very rapidly, and the cost of their fattening can be lessened considerably by mixing boiled potatoes with their barley-meal at first, or pollard, to reduce its cost, merely finishing up with barley-meal alone. A large hog of the improved Berkshire breed has been thoroughly fattened upon two sacks only of barley meal, which has been eked out by the aids mentioned.

When the pig's consumption of food begins to fall off, he will not pay for any further feeding, and ought then to be disposed of. The mention of potatoes reminds us that when food of a heating nature has been given, such as beans, an eruption will sometimes make its appearance upon the ears, and this will get removed by a mixture of potatoes with rather more salt than is generally given, pigs needing a certain portion of salt in all their food. Some persons find it advisable to use a mixture of peas, Indian-corn, and barley-meal, instead of barley-meal alone, which lessens the expense when fattening.

Potatoes, indeed, have for a long time been an important ingredient in the fattening of pigs, and when mixed with a moderate portion of meal, forms an economical food of great value, while bean-meal, pea-meal, Indian-corn-meal, and oatmeal very nearly approach barley-meal in value as food for pigs.

Oatmeal mixed with skimmed milk or butter milk is said to produce the most delicate flavoured bacon, and is extremely nourishing.

In fattening for bacon, large pigs of good age and breed should be preferred, and they should be at least fifteen months old to make first rate breeders. Young pigs require good nutritious food, while old pigs will fatten upon almost anything, when they possess a fair amount of good lean flesh to begin upon.

Pigs fatten quickly when animal matter is given to them, such as grease, greave cakes, &c., but it makes the flesh rank in flavour ; and it is not advisable to let them have such kinds of food.

At the same time a change of victual is advantageous, and a little salt should be frequently used ; the main thing in fattening pigs is to give the meals with the greatest regularity, and in the sweetest and most attractive form—at least three times a day. The pigs should have enough, but not so much as to be wasteful, and the troughs cleaned, or, better still, washed out daily.

In hot weather some persons make a practice of throwing water over their pigs, with the view of cooling both the pigs and the sty.

FEEDING FOR SHOWS.

Everyone knows that pigs are, when exhibited at the various shows and Agricultural meetings, of enormous weight and fatness. The Smithfield Club has a standing rule which directs all exhibitors to state how their animals have been fed, and the chief food used is found to be barley, bean, and pea-meal, peas, Indian-corn-meal, potatoes, middlings, coarse flour (not the inferior barley-meal, or pollard which passes under that name) skimmed milk, whey, &c. Barley and pea-meal generally carries the palm as being the choicest food, and the most effective in its operation of fattening.

WEIGHTS AND MEASUREMENT.

It is very commonly calculated by salesmen that the dead weight of an animal is one half of that which it weighs alive. This is a very convenient rule for the purchaser to go by, but not an advantageous one for the seller, as, in most cases it will be found to be nearer three-fifths than one half. A well-known stock bailiff of considerable experience always calculated that the dead weight was equal to $\frac{550}{1000}$, that is to say, about eleven twentieths of the live weight.

The difficulty found in correctly ascertaining weight, has led to the measurement of live stock, and tables have been constructed by several ingenious persons, by which weight can be calculated according to the animal's dimensions.

This is now always done in the case of cattle, and is considered equally applicable to every kind of animal, the process being conducted in the following manner: The girth is taken by passing a tape measure or cord just behind the shoulder-blade and under the forelegs, this gives the circumference; and the length is taken along the back from the foremost corner of the blade bone of the shoulder in a straight line to the hindmost point of the rump, or to that bone of the tail which plumbs the line with the hinder part of the bullock.

The weights stated in three tables, published by Renton, Oary, and M'Derment, as will be seen by the following, nearly all agree with each other, and having been tested by animals measured when alive, and afterwards killed and weighed, they were found to approximate so nearly to the truth as to afford a very accurate rule:

GIRTH.	LENGTH.	RENTON'S TABLE.	M'DERMENT'S TABLE.	CART'S GAUGE.
ft. in.	ft. in.	st. lb.	st. lb.	st.
5 0	3 6	21 0	20 11	21
	4 0	24 0	23 11	24
5 6	3 9	27 1	27 0	27
	4 9	34 4	34 2	34½
6 0	4 6	38 8	38 8	38¾
	5 0	43 1	42 12	43
6 0	4 6	45 9	45 3	45½
	4 9	48 0	47 10	48
7 0	5 6	64 6	64 2	64½
	6 0	70 5	69 13	70¾
8 0	6 6	99 8	99 0	99½
	7 0	107 5	106 9	107½

The tables are calculated upon the stone of 14lb. avoirdupois, by multiplying the square of the girth by the length, and this product by a decimal which may be assumed as nearly ·238, for the live weight. The dead weight is ascertained by multiplying the live weight by the decimal ·605; thus $\frac{605}{1000}$ will give the product of the four quarters.

M'Derment suggests that in the case of very fat animals, one-

eighteenth or one-twentieth part should be added to the weight
obtained by measurement; and below the ordinary state of fatness the same proportion should be deducted; also that old
milch cows, which have had a number of calves, should have one-
ninth or one-tenth of their weight deducted.

Cary's gauge is an instrument made in the form, and on the
principle, of a slider rule, giving the weights marked in stones of
8lb. and 14lb.

No one, however, who fattens hogs on a large scale, or extensive
grazier, should be without a steelyard; for, by its constant use
while the animals are fattening, he can instantly ascertain the
state of progress of the beasts, and thus be enabled to compare
together their expense and their improvement, although this
cannot be done with accuracy unless the animal has fasted for at
least twelve hours.

KILLING.

When it is intended to kill a pig, due notice should be given
to the man who feeds it, in order that the animal designed
for slaughter be kept without food for twenty-four hours, so
that the intestines are well emptied, but it should be allowed
water. The most common practice is by "sticking" in the
neck, and a pork butcher, who has had a great deal of experience, will deprive them of life almost immediately; but many
adopt the method followed in killing oxen with the poleaxe, and
use a kind of hammer, three feet in length, with a spike about
three inches long at the head.

This is struck with a firm blow on the low part of the forehead, so as to immediately enter the brain, killing the pig
instantaneously. The vein in the neck is then quickly opened
by the knife, so as to allow the blood to flow out freely, which is
carefully saved, and often profitably made use of.

SCALDING AND SINGEING.

When pigs are scalded after being killed, for the purpose of
removing the hair, they are placed upon a bench, or boards put

upon trestles, and scalding water is thrown over the carcase. Some persons use a shallow tub—often a mashing tub—but this process is apt to wet the carcase too much, the only object being to take off the hair, which scrapes off readily with the knife when effectually scalded.

In many counties, as Hampshire and Berkshire, the common practice is to singe off the hair, which is called "swaleing," which is done immediately the animal is killed. The carcase is laid on the ground on its side, and a thin covering of fresh dry straw placed over it. This is set on fire, and renewed as often as necessary, care being used not to raise too fierce a flame so as to scorch the skin. When one side is thoroughly done, it is then turned over, and the reverse side served in the same manner.

In the counties where this practice is followed it is considered a far better one than scalding, which is thought to soften the rind and injure the firmness of the fat. When thoroughly singed, all the burnt bristles and bits of dirt and ashes accumulated in the process are scraped off thoroughly dry.

Cutting Up.

Henderson recommends the following method of cutting up, &c.: "After the carcase has hung all night, it should be laid on its back on a strong table. The head should then be cut off close by the ears, and the hinder feet so far below the houghs as not to disfigure the hams, and leave room sufficient to hang them up by; after which the carcase is divided into equal moieties, up the middle of the back bone, with a cleaving knife, and, if necessary, with a hand mallet. Then cut the ham from the side, by the second joint of the backbone—which will appear on dividing the carcase, and dress the ham by paring a little off the flank, or skinny part, so as to shape it with a half round point, clearing off any top fat that may appear. The curer will next cut off the sharp edge along the backbone with a knife and mallet, and slice off the first ribs next the shoulder where he will find a bloody vein, which must be taken out, for, if left in, that part is apt to spoil.   The corners should be squared off when the ham is cut out.

"When this is done, give each 'flitch '—as the sides are called —a powdering of saltpetre, and then cover them with salt, and let them remain in a cool place, and proceed in the same manner with the hams.   In this state they may lie about a week, after which they should be turned and fresh salted, and in two or three weeks longer they may be hung up to dry in the smoke house ; but if allowed to remain for a month or two until it may be convenient to dry them, no harm will occur, provided they be occasionally turned."

While the salting is being done, as much of the salt will melt, it is expedient to place the pieces upon a sloping board, so that the liquid may run off.   They should also be carefully inspected from time to time, and salt rubbed in with the hand on any places which may appear to need it.   While passing through the process of salting the rind should be underneath.

Smoking.—In old-fashioned farmhouses, where wide chimneys are to be found and wood is burned, the flitches and hams used to be hung up to catch the smoke from the fire ; but a better plan is—where any quantity of pigs are killed—to have a separate smoking house, which can be made at a trifling expense out of a few boards, as there the operation can be performed in an equal and even manner.   About seven feet high is sufficient, closed on all sides, with merely the door for entrance, and a small hole at the top for the smoke to pass through after having performed its office in smoking the bacon.   The flitches and hams should be well rubbed over with bran and hung up on stout cross pieces about three feet from the floor.   They may hang as thickly as possible, so long as they are not allowed to touch each other. The necks of the flitches should hang downwards.   The floor should be covered with sawdust to the depth of five or six inches, and then kindled, when it will smoulder without flame.   Flitches will be mostly cured in ten days or a fortnight ; but the hams, being thicker, require longer time.   A great number of hogs can be smoked in this way in a very short time.   In the Peninsula, and on other parts of the Continent, where hams are cured in an excellent manner, it is usual to make a mixture of sugar, saltpetre, and salt, in the proportions of three pounds of salt to one of sugar, and two ounces of saltpetre. The sugar causes the fibres of the meat to be mellow, and removes the harsh pungency of flavour which is often communicated to bacon and hams from the too liberal use of salt. The Westphalian hams are mostly cured in this way.

RINGING.

Ringing has been stigmatised as a cruel practice, but we are unable to see how it can very well be dispensed with. The amount of mischief pigs can do when their snouts are unrung, by ploughing up a meadow which is soft, is something surprising to those who have seen the grass cut up by them, where they have made holes almost large enough to bury themselves in, as well as forming long furrows in various directions.

To prevent this, some persons cut a slit in the nose, but, in the first place, it is extremely doubtful whether it answers the desired end; and, in the next, it gives the head a very ugly appearance.

In order that the operation should be as little painful as possible, the young pigs should be rung as soon as convenient after they have been weaned; it is easier for the operator, and is altogether a less formidable transaction than when deferred later on.

Even if they are not turned out in the fields, it will prevent them overturning their troughs and "rooting" about their sties, while it is the means of preventing their making an attack upon one another, which they will do occasionally when their hard noses are left intact.

The simplest and readiest system of ringing is to cut up some sufficiently stout wire for the purpose into lengths about four inches long, and push the wire—one end of which has been sharpened—half its length through the nose. The ends are then brought together and twisted with a pair of pliers, or pincers, a small bit of stick being placed between the loop, so as not to screw it up tightly enough to hurt the pig.

For old sows, and larger pigs a somewhat stronger ring is required, there being several kinds in use ; and, as these want renewing and come out occasionally, the larger animals are obliged to have a cord placed round the upper jaw, and be tied to a post during the operation to hold them securely, to which they have a most decided objection.

If the wire is put too far in the flesh of the snout, it is not unlikely to cause a complaint which is known as "the snuffles," causing the animal to breathe with a loud noise. A pig so affected is said not to fatten so well, and therefore care is necessary to place the wire near the tip of the nose, but having sufficient hold of it to allow it to remain where it is placed.

This operation should never be performed upon sows that are with young.

CHAPTER III.

SHEEP.

*Sheep Farming—Varieties of Sheep—Original Breeds of Sheep—
Short Wool—Long Wool—Shetland Sheep—Welsh Sheep—
Advice on Buying—Accommodation for Sheep Folding—Sheep-
cotes—Wintering Sheep in Fold—Yards—Respective Advan-
tages and Disadvantages of Various Accommodation—Labour
Required for Superintending Sheep—The Shepherd—The
Shepherd's Dog—Feeding—Summer and Winter Feeding—
Fattening for Market.*

SHEEP FARMING.

SHEEP farming is found to be very profitable, when it is conducted
in a thorough and proper manner, owing to the comparatively
high price wool fetches in the market for manufacturing purposes, and the value of the mutton, for which there is a constant
demand; and the obvious course is, to have a breed of animals
which unite in their bodies wool-producing and meat-producing
qualities. There is an immense variety of different breeds of
sheep in England, so that there is not the slightest difficulty in
obtaining a race of animals precisely suited to the pasture
which may be waiting to receive them, whether it be that of
a rich, alluvial plain, or a mountainous district for animals to
pick a living from heathy herbage with which the hills may
be clothed.

The old breeds of sheep did not combine the various qualities
that are looked for in the present day—viz., plenty of wool on
their backs, weight of carcase and good quality of flesh, and
aptitude to feed.

VARIETIES OF SHEEP.

Original Breeds of Sheep.—The original breeds used to be divided into two classes, termed " short wool " and "long wool." The animals fed upon mountainous pastures were small in size, and covered with a close but short coat of fine wool, which earned for them the distinctive term of " short-woolled," in contradistinction to those fed on the rich low lands or fat marshes, which attained larger size and greater amount of fat and flesh, together with longer and coarser wool, which were termed "long woolled." These were the two broad distinctions that used to be drawn between existing species of sheep ; and what are now termed the old-fashioned breeds, have been tabulated as under, with the weight of wool of each, and the dead weight of the flesh per quarter :

SHORT WOOL.

	WEIGHT OF FLEECE IN THE YOLK AND UNSMEARED.		DEAD WEIGHT OF THE FLESH PER QUARTER.	
	lb.	lb.	lb.	lb.
Southdown polled	2½ to	3	18 to	20
Wilts and Chiltern horned	2 ,,	2½	14 ,,	18
* Dorset ditto	3¼ ,,	3½	16 ,,	20
Portland ditto	1½ ,,	2	8 ,,	10
Exmoor and Dartmoor ditto	3 ,,	4	10 ,,	12
Cornish ditto	2 ,,	2½	12 ,,	15
Ryeland polled	1¾ ,,	2½	13 ,,	16
Dean Forest and Mendip horned	1½ ,,	2	12 ,,	14
Norfolk ditto	1¼ ,,	2½	14 ,,	18
Cannock Heath polled	2½ ,,	3	16 ,,	20
Shropshire Morf horned	1¾ ,,	2	9 ,,	13
Delamere Forest ditto	1¾ ,,	2	8 ,,	10
Herdwick polled	1½ ,,	1¾	9 ,,	12
Cheviot ditto	2½ ,,	3½	12 ,,	18
Scotch Heath horned	2½ ,,	3	13 ,,	16
Shetland horned and polled	1½ ,,	2	8 ,,	9
Welsh Mountain ditto	3 ,,	2½	10 ,,	14
Pure Merino, horned and polled	4 ,,	5	15 ,,	18

* *Dorset Sheep.*—The fact of the Dorset receiving the male at an earlier season than other British sheep causes them to be used for rearing early lambs for the London market, which are sold about Christmas time and January. At Weyhill and other fairs in Wiltshire and Hampshire, farmers resort to purchase ewes in lamb, with the view of fattening the lambs first and the ewes afterwards. The earliest lambs slaughtered previous to Christmas are most of them bred in the house—hence the term "house-lamb."

LONG WOOL.

	WEIGHT OF FLEECE IN THE YOLK AND UNSMEARED.		DEAD WEIGHT OF THE FLESH PER QUARTER.	
	lb.	lb.	lb.	lb.
Bampton, Notts, polled	7 to	8	22 to	28
South Ham ditto	8 „	9	18 „	22
Cotswold ditto	7 „	8	26 „	34
Romney Marsh ditto	6½ „	8	22 „	28
Dishley ditto	6 „	7	21 „	25
Leicester and Lincoln ditto	8 „	10	24 „	32
Teeswater ditto	7 „	8	26 „	36
Irish	6½ „	7½	22 „	26

The above table displays the average weight at two years to thirty months old of both ewes and wether ; and in Cotswolds, Leicesters, and Lincolns, were found both wool and meat; but their quality was so inferior as to be only saleable in the poorest markets, such as manufacturing centres; while the Southdown gave the very best quality of meat with but a small weight of wool. It was, therefore, desirable to have a cross, such as Shropshire or the Oxfordshire, with the Down Leicester sheep, which will be found to combine the qualities wanted.

The Shropshire sheep, as a cross, have been found very valuable in some of the western midland counties; while the Oxford, possessing characteristics of the long-wool species, are very suitable to the arable soils of that county, and were originally derived from a cross of the Cotswold and Hampshire Down. They are hardy and moderately prolific, and come to maturity early, coming out at twelve to fourteen months old with 7lb. to 8lb. of wool, and weighing 20lb. per quarter. The old Norfolk black-faced sheep, which gave but a small yield of wool, are now seldom seen, but traces of their breed may be easily recognised in many cross-bred sheep that are commonly found in the eastern counties. These traces may be identified in the qualities of hardihood, large frame, and strength of constitution, combined with profuse milking powers. This original stock, grafted with Cotswold, improved Lincoln, or Leicester sheep (the latter improved from the original), produces valuable animals.

There are two kinds of sheep which are generally regarded as aboriginal races—the Shetland and Welsh sheep. The former has some peculiar characteristics, being capable of supporting extreme cold and hunger, and in winter feed upon seaweed, their instinct communicating to them the first ebbing of the tide, when they may be seen leaving the hills for the sea-shore. It is said they will eat animal food, and have been fed in seasons of scarcity upon dried fish. The fleece is composed of an outer coat of long hair, which is termed "scudda," which projects from the wool and throws off the wet, being a protection bestowed by nature against the furious storms which rage in these islands at times, from which there is but little protection. At the beginning of the summer the true fleece becomes detached from the skin, and if not collected in time, rises through the hair and falls off. To prevent the loss of the wool, at the proper time the sheep are gathered together, and instead of being shorn, the wool is pulled off by the hand. The wool of each sheep when thus separated from the "scudda" weighs from one and a half to two pounds, and forms the well-known Shetland wool, which is used in making the fine shawls known as Sheltand shawls, worn by ladies, and the better known Shetland hose.

There is a certain resemblance between the fine Welsh sheep and the Shetland, though each possess distinct features. The former have long tails reaching below the hocks, while the Shetland have a short, broad tail. Welsh sheep are divided into classes, as, the sheep of the higher mountains, the soft-woolled sheep, or sheep of the Welsh hills, and the larger species of the low country. The latter, however, are not really Welsh, being either the smaller kinds of English sheep, or crosses from them upon the old stock. The soft-woolled sheep is the race which furnishes the wool employed in making Welsh flannels, and the mutton used to be sent largely to London, where it formerly fetched a higher price than English mutton, but now their relative positions are reversed.

ADVICE ON BUYING.

From the short sketch we have given of the different breeds and crosses of sheep, sufficient may be gathered to guide the intending purchaser as to the proper selection he should make

and his choice should be mainly influenced by the character of
his pasture, or, if his land is chiefly arable, by what green crops
he can grow to the greatest advantage, so as to chime in with
his general arrangements in the best manner.

In England the breed most in repute is that between Southdown ewes and long-woolled rams. In Scotland, it is most frequently considered that a cross between Cheviot, or black-faced
ewes, and Leicester rams, best suits the general routine of sheep-
farming there, and the natural circumstances appertaining to the
country. The main point, however, is, at the first start off, to
purchase a breed adapted for the situation, land, and general
capabilities of the farm.

ACCOMMODATION FOR SHEEP.

Folding.—The folding system answers best when a variety of
successive crops are produced, and some flock-masters recommend two or three changes a day. Shepherds, however, often
object to the trouble of this. In the large folds common in
Lincolnshire a good deal of food is trodden down, and the plan
is wasteful. Some of the long-woolled sheep are bad travellers,
and experience bad effects from being driven about, and when
made to undergo much exertion suffer severely. Folding has
been approved now for a very long time, and in the "Survey
of Somersetshire" Mr. Billingsby says: "In a rich, fertile
country, where the quantity of arable land is small, and in mere
subserviency to the grazing system, where dung is plentiful,
and can be put in the corn field at a small expense, and where
each sheep is highly fed, it is not to be wondered that the
folding system should be held in derision and contempt; but I
will be bold enough to repeat that in a poor, exposed, and
extensive corn farm, the soil of which is light and stony, it is
the *sine quâ non* of good husbandry. Let me ask its opponents
whether the downs of Wilts and Dorset would wave with
luxuriant corn if folding were abolished? No. The farmer
would plough and sow to little purpose were his fallows to
remain untrod with the feet, and unmanured by the dung and

perspiration of these useful animals. Besides, in the hot summer months, nothing is so grateful to the flock itself as fresh ploughed ground; and sheep will, of their own accord, retire to it when their hunger is satisfied." Sheep farming has, indeed, of late years, effected a wonderful revolution in some parts of England, and has raised the annual value of the land to three times that of its original worth.

On small farms, where the number of sheep kept is too small to require the constant services of a shepherd, the manure may be secured for the arable land by having a permanent fold adjoining the farmyard, on a dry situation—the walls of the yard and buildings affording shelter—in which large quantities of manure will be collected with but little trouble. The fold should be well littered with leaves, stubble raked off land, or refuse straw. The animals can then be attended to with comparatively little trouble, and all driving the sheep about will be avoided, except when it is deemed necessary to put them on the pastures. The manure produced in this way will be found extremely valuable. Wethers can be fattened for the market and the ewes, at lambing time, can be provided with shelter. As many divisions as may be wanted can easily be made with thatched hurdles, which are most useful contrivances in the hands of a clever man, and they are not so hot as enclosures with solid walls.

Sheep-cotes in exposed situations can also be made from these, as well as stells intended for refuge in sudden snow storms in mountainous districts. The latter are made in various shapes to guard against the effects of the tempest, and for preventing the snow from lodging within them, sometimes in the forms of the letters T, H, or S, or circular in shape.

Wintering Sheep in Fold Yards.—Sheep are liable to foot-rot when wintered in fold yards, which arises from the hoof constantly growing and not receiving the wearing action to which it is subject when the animal is in its unconfined state, and, in consequence, it curls inwards and lameness ensues. To prevent this, the feet should be carefully pared upon first being put into the fold, and looked to every three weeks or so afterwards.

It is important that there should always be sufficient litter of some sort to keep the feet of the sheep from coming in too close contact with the heating manure.

Advantages and Disadvantages of Various Accommodation. —The advantages and disadvantages of different methods of folding sheep stand thus : The ordinary movable fold permits of the land being regularly manured by the sheep without any cost beyond that of the labour incurred in moving the hurdles. As, however, they are ordinarily pitched upon arable land the dirt and wet are sometimes injurious to the animals, while, if placed at a long distance from the pasture, the extra labour incurred in the journeys prevents the stock from fattening.

The "stell" is only resorted to as a means of security and shelter in districts which are liable to the visitation of mountain snow-storms.

The "standing fold," which is placed in a dry spot, in the most convenient situation, and bedded either with sand or litter, supplies an accumulation of manure which can be applied wherever it is wanted, at the right time and proper season ; but, on the other hand, the soil loses the supposed advantages of the teathe, while the expense of mixing the compost and spreading it on the land has to be considered.

In the case of the "cote," all the advantages and disadvantages of the several systems are combined, with the addition of shelter, but the cost of erecting sheds is to be considered, and if constantly used, they make the sheep very tender, and their health often suffers in consequence.

Protection from extremes of heat and cold is very necessary, and in inclement winters, especially upon clay soils, there is an advantage in feeding sheep under the cover of well-ventilated sheds; but care must be taken to prevent them from getting over-heated, and it is sometimes found necessary to shear them in order to prevent this.

Like the man who was the guest of the satyr in La Fontaine's fable, who blew upon his finger ends to make them warm, and upon his porridge to make it cool, the fleece of the sheep possessing a non-conducting property, it will keep in heat, or keep it out, and excepting in extreme cases, their natural covering is a sufficient defence against cold and wet.

In experiments made in feeding in sheds it has been found that the sheep made a quicker relative progress during the first six or eight weeks of their being housed than when confined for a longer period, a result that would seem to point to the time most likely to be of advantage with animals nearly ready for market.

LABOUR REQUIRED FOR SUPERINTENDING SHEEP.

One man and a good dog can look after a great many sheep; but it is highly important that all who have sheep put under their care should have acquired the duties of attending to them whilst young, and be of kind and considerate disposition.

Persons who have not had that experience in the ways of sheep, are, unfortunately, apt to display acts of unnecessary violence and senseless anger, and knock the poor beasts about in a cruel way. In Spain the shepherds have an excellent plan of teaching a few wethers to obey their call and follow them, so that the whole flock may be led anywhere with the greatest docility.

Sheep are often supposed to be somewhat stupid and not very teachable, but it is really not so. They can be easily made to answer to their names, and any sheep can be taught to do this by being rewarded with a handful of salt, or by feeding it with the hand when a lamb.

The Shepherd.—Very much depends on the shepherd in the success of sheep-farming. He should not only take an interest in the animals he has to look after, but be sufficiently acquainted and instructed in the common surgical operations, and modes of prevention of disease attendant upon the care of a flock.

Long experience can only make him acquainted with their habits, and in some families, where sheep are largely reared and fed in the neighbourhood where they reside, the occupation is hereditary.

Although there are long periods of daily leisure, when the shepherd appears to have nothing to do, but to see his flock grazing, there are often very arduous duties to perform, particularly in mountainous districts in stormy, snowy weather, and these often have to be done at all hours of the night, for a good servant knows the property under his care is extremely valuable and demands constant attention.

A calm-tempered man makes the best shepherd, one who is honest, active, and careful, and takes a real pride and interest in his sheep.   It is a good plan to allow the shepherd a certain premium on the number of lambs reared, as this makes him an active working partner in the business, as it were ; but it is a very bad one to allow him to have a certain portion of the flock to himself, on paying a portion of the expense of their feed, as is sometimes done.   The system leads to petty frauds and temptations to dishonesty, the shepherd's stock being generally the best, and suffering from fewer casualties than the remainder. It is unfortunate that this should be the case, but it will ever remain so while human nature is as it is.   Even with an honest desire to do that which is fair and right, one cannot help feeling a greater involuntary amount of interest in what is one's own than in another's.   There are many excellent employers of farm labourers who will not allow their men to keep pigs on this account, as they consider they give rise to a temptation to steal the corn and other provender about the place.

It has been estimated that one experienced man solely engaged in following the duties of a shepherd can, in the summer months, under common grazing in pastures or clover, attend to about 800 sheep of an average flock.   In the case of a breeding flock the number would be somewhat less, but in that of a fattening flock somewhat greater.

In the winter months he would have his hands full, to attend to five hundred, fed upon cole seed and turnips—the sheep feeding themselves, the shepherd only dragging out the hulls of the turnips as it becomes necessary to do so ; but if the sheep are fed wholly on turnips, four hundred would be a fair average to engage one man's full attention.   If the sheep are fed with cut turnips in troughs (and this is the proper way of feeding them to the best advantage), two hundred sheep are as many as one man can well attend to, for, in addition to moving the hurdles, he has to pull up the turnips, take them to the cutter, clean them, and carry them to the troughs. If the shepherd's wages are a pound a week, it will thus, in the winter, cost 10s. per week a hundred to look after sheep, though cheaper labour is to be obtained in the winter, as any ordinary labourer can cut turnips and feed the sheep, the shepherd superintending the whole and attending to other parts of his flock in different places on the farm.

During the summer months the shepherd lends his hand in making hay, thatching ricks, and other jobs which he can find time to attend to.

One experienced agriculturist estimates that when the turnips are heaped up and covered with mould, as they are sometimes stored, with a little straw under the mould, it takes a man and a boy to attend to one hundred fattening sheep and do the routine work of uncovering the heaps, top and tail the turnips, clean and cut up the roots, fodder with hay, and perhaps corn or cake, twice a day, and move the racks, troughs, and hurdles each day, as may be required.

The same authority estimates that the same labour would be adequate for attending to 250 sheep which eat the roots off the land, the hulls to be picked up each day, hay supplied as above, and the hurdles moved when necessary, and that one good shepherd can keep from 250 to 300 breeding ewes, making out all the sheep when sold to be in fat condition.

The Shepherd's Dog.—There are two breeds of sheep dogs commonly used, the one chiefly seen in England, with whose sight most persons are familiar, with his cropped tail, and the shag-haired "colley," as he is termed, in the more northern districts of the kingdom, where there are many extensive sheep walks.

Both breeds are generally considered on a par and of equal sagacity, not only readily obeying the commands of their master, but at times, when occasion demands, acting spontaneously for the security and well-being of the flock, which renders him a valuable acquisition, and, it is needless to say, saves the legs of the shepherd, who soon would not have one to stand upon himself if he had to run after his charges upon the various occasions when they are tempted to stray.

One fault in their education is, that in droving they are generally taught to bite the sheep, and although not in such a manner as seriously to injure them, yet heavy ewes sometimes slip their lambs through fright from this cause.

The dogs in Spain do not bite, and so far from being objects of terror to the sheep, at times of impending danger, instead of shunning them, the sheep will gather round them for protection.

FEEDING.

There are two methods commonly followed of feeding sheep, the one being to eat the roots off the field where they are grown, if the land is dry enough to put the sheep with safety upon it, or to cart the turnips to a bare pasture, where they are sliced and given to the sheep. In the first method the field is hurdled off in partitions. When the better part is eaten off, the remainder of the roots are loosened from the soil by a gutter. The lambs should on no account be allowed to touch clover or turnips when the frost is on them.

The practice which used to prevail of giving breeding ewes a large quantity of roots is pronounced to be a bad one, and it is said that a good turnip year in Norfolk used to be followed invariably by a bad lambing season. Straw and hay-chaff, made palatable by a little artificial food and a moderate quantity of roots, is considered far better management than used to prevail under the old system.

A mixture of beans or peas, barley, wheat or oats, and palm-nut meal, is considered to make the best mixture. Lincolnshire sheep growers frequently give their ewes from $\frac{1}{2}$lb. to 1lb. of linseed cake daily during winter, considering they are repaid for the outlay in the increased quantity of wool, value of manure, and healthy condition of the animal.

At times there are great losses both in ewes and lambs, which may often be traced to wet lairs and insufficient feeding, which plant the seeds of consumption in sheep.

The adoption of a straw yard at night, is considered a good one. The sheep, lying warm, will not eat so much food ; and, if kept in when the ground is frosty, and prevented from eating the frozen grass, scouring is avoided. Baked chalk, in powder and lumps, placed in the feeding troughs, is a good safeguard against looseness in lambs. When lambs are allowed to range, shelter should be provided in the form of a few thatched hurdles properly fixed, or an old waggon half laden with straw.

If dry food can be given to sheep in the fold—either chaff of oat straw, pea haulm, or hay—it will be a good preventative against their being "hoved," and wards off any tendency to looseness ; but, whatever is given to them should be given with regularity. Roots should never be given till quite ripe. When sheep are first put upon swedes, serious losses often happen from the roots being unripe, and consequently deficient in sugar, so that the nitrogenous matter, probably combined in an unhealthy form, irritates the bowels. Locust beans will stop this scour, and they are very valuable on this account ; but it is best to avoid occasion for the remedy.

The chief object aimed at in feeding sheep is, of course, to get them fat, and out of hand ready for the butcher as soon as possible. The older the mutton, the finer the flavour, but there is very little old mutton to be had nowadays. It is said by connoisseurs that a spayed ewe kept for five years before she is fattened, produces superior mutton to that of any wether. Of course, it would be highly unprofitable to keep such an animal.

In the summer time, when there is not land enough for them to feed in a natural manner, it has become the custom to soil them with various artificial grasses, and if they are then not ready for market at the close of the season, they are finished off with swedes, or hay and oilcake.

Nothing has a greater effect upon the increase of flesh than oilcake, for if even the sheep have been fed upon turnips, 1lb. of oilcake per day has been found more profitable, when purchased at £8 per ton, than turnips at £4 per acre. .

When turnips are drawn, it is a good plan to keep them a day or two before they are given to the sheep, in order to allow their watery juices to evaporate somewhat, by which means they become more wholesome and nutritive.

With whatever roots they may be fed upon, hay or straw should always be given to promote digestion—whether mangold, cabbages, swedes, carrots, or parsnips, upon which their progress will be found nearly equal. Parsnips are eminently nutritive, and improve the flavour of the meat, but they are not nearly so largely employed as they might be.

Turnips are the stock feed relied upon for sheep, one experienced manager stating that he never found his flocks thrive so fast on any food as swedes, with a little freshly thrashed barley straw. When he gave them good green hay, instead of straw, he found they ate more hay, and less turnips, and in consequence did not thrive so fast; the more turnips they consumed in a given time, the quicker they became fat, the dry food being only required to assist the digestive organs, and to correct the watery properties of the turnips. Straw, being less costly than hay, can be advantageously used for the same purpose, it being a well ascertained fact that hay or straw promotes digestion.

Fattening for Market.—The chief object, as mentioned before, in feeding sheep, is to get them ready as quickly as possible for market; and in converting the crops of a farm into wool and mutton, it must be ever borne in mind that, to do this in the best and most economical manner, the animals must be so treated that their uninterrupted progress is insured. A check in their growth involves both a loss of time and a waste of food, and the ultimate result is affected unfavourably.

There is a certain degree of fatness which is necessary for the consumer, and profitable to the stock owner; but, when carried too far, it is injurious to the quality of the meat, without adding to the gain of the feeder, though the increase of tallow puts money into the pocket of the butcher.

Oilcake, if used too freely, imparts an oily softness, as well as a yellow tinge to the fat; and to delicate tastes, the lean is also objectionable. The quantity of inside fat depends largely upon the age and time of fattening, accumulating much more in old sheep than young ones. The tallow of a wether under ordinary management generally averages from an eighth to a tenth of its dead weight, though animals have been known to have had as high a proportion as one-sixth loose fat.

Oilcake, linseed, and other artificial foods of a similar nature, are used for creating flesh quickly, often with profit, but it must be admitted to the disadvantage of the mutton ; but many sheep are fattened upon the rich grazing lands of the north of England, in Ireland, and the marshes of Essex, Somerset, and Kent, where they gradually attain a state of perfection, and this is usually looked upon as the preferable mode of feeding, but then it is not every kind of pasture which is capable of fattening the animals, although it may support them in health.

Corn and pulse are considered to be the most efficient food in fattening all cattle and giving firmness to their flesh and tallow ; but the consideration for the grazier is to effect that result in the most economical manner, and it is very questionable whether grain can, in this country, be profitably given to sheep. It can, doubtless, be used to advantage when the season has been such as to render barley unfit for malting, and damaged oats, together with potatoes, in the proportion of one-third of the former to two-thirds of the latter, has been used with good effect, but bruised oilcake is more commonly resorted to, as a constant supply may be always reckoned upon.

A good mode of giving potatoes to sheep in the depth of winter is to take the sheep into the foldyard, when the potatoes are cut into slices, and placed in troughs under the shelter of the sheds. Sheep have been found to fatten upon these very quickly.

Of the value of chaffed straw mixed with roots too much can scarcely be said, both in fattening as well as in the course of ordinary feeding.

An able advocate (Mr. Coleman) for the employment of more straw and straw-chaff on the score of economy in feeding stock, in his prize essay on the "Management of Sheep Stock," pub-

E

lished in the *Journal of the Royal Agricultural Society*, says:
"Hitherto farmers have supposed that a bellyful of turnips was
necessary for a breeding animal, and have based their calculations
on their stock of roots that were to be thus wasted. The past
winter has taught us to give these roots in a healthier form, eking
out the supply by a nice admixture of other food, supplied in a
palatable form. I have lately inspected a flock of Hampshire
Down ewes that did not have a root before lambing. They ran on
grass land during the day, being hurdled at night, so as to dress
and improve the pasture. Morning and night they got trough-
food, consisting of straw and hay-chaff—two-thirds of the former
and one-third of the latter—bruised oats, and palm-nut meal.
The cost of the artificial food amounted to 2¼d. a head weekly.
Not one ewe died during the winter, and I never saw animals in
a more promising state for lambing."

And, again: "The object of this essay is to point out the best
means of increasing sheep stock. Here, then, is one way. We
must make one acre of turnips keep twice as many sheep as
hitherto in a far more healthy condition. Last winter, in too
many cases, the difficulty was to find any roots at all; but great
and lasting good may be anticipated from the evil then felt. I
saw many flocks during the past winter living on damp chaff,
with a little artificial food, and doing as well as could be wished,
with every prospect of a healthy produce and plenty of milk. I
have long desired to see an economical plan of pulping roots
devised, as the animal might then be induced to eat a large
quantity of straw-chaff, rendered palatable and nutritious by a
small addition of artificial food. Nor would such a system be so
extravagant as at first it might appear. Let us assume, by way
of example, that one crop of turnips equals fifteen tons per acre,
and that, instead of 20lb. per head, we give 10lb. (amply sufficient),
with 1lb. of straw-chaff and ¼lb. per day each of artificial food,
and it follows that 100 sheep will consume an acre in thirty-
three days, and 7cwt. of extra food will be spent on each acre,
besides 1½ tons of straw, so as to considerably increase our produce of corn, besides the chief object of keeping a heavier stock
of breeding sheep in a healthy state." At one of the Royal Agri-

cultural Society's discussions, the same gentleman whom we have quoted drew the attention of those present to the more economical system of feeding sheep with straw-chaff, so as to increase the returns, and leave the land in a better condition for corn, his recommendations being, as above stated, to reduce the quantity of roots, and use more dry food, as straw, in combination with a small quantity of artificial food to act as a stimulus to digestion.

A breeding ewe will consume one-fourth of its live weight of turnips, or 20lb. to 30lb. per day, of which nine-tenths are water, so that if the roots were reduced one-half, and an equivalent substituted in the form of straw and condimental food, at the same time attending to the external comfort of the animal, a great point would be gained.

From personal experience, Mr. Ooleman was convinced that good straw may be economically substituted for hay in the winter feeding of sheep, even without any artificial food, though the cost of the latter is so slight that it can be very advantageously given, and that by giving sheep partly straw fodder, and partly roots, while feeding on the land, the value of the manure left would be increased by more than one-half.

CHAPTER IV.

SHEEP (CONTINUED).

Washing—Shearing—Lice and Ticks—Marking—Markets for Wool—Salving—Breeding—Age and Period of Gestation—Improvement of Breeds—Age of Rams—Influence of Sire and Dam—Flock for Breeding Purposes—Yeaning—Weaning—Age of Sheep—Prime Mutton—Descriptive Terms.

WASHING.

IN order to disencumber the fleece of any coarse dirt which may have been deposited in it, it is usual to wash the sheep previous to shearing. The common practice for this purpose is to form pens, or artificial pools, railed round with one rail only, just above the water, so that the sheep may be thrust under. A long pole, called "a poy," is employed for this purpose. This pole has a projection of about 6in. on each side of the end, so as to enable a man either to pull the sheep to him or push it from him, as may be necessary.

Where there are small rivulets, 2ft. or 3ft. in depth, these are generally chosen, with the wash-dyke pointing against the current, so that the foul water may be kept draining from the sheep. Or, a dam is placed in a convenient place, with a flood-gate in the middle, by which means a pond of water may be formed at pleasure, and let off as needful, a pen being formed on one side for washing, and on the other a path is hurdled off for the sheep to ascend when the washing is done.

The business is generally made a point of being got over each day by two or three o'clock in the afternoon, so as to give the sheep an opportunity of getting somewhat dry, and get back their natural warmth before sundown, and the approach of the evening damp.

Some persons repeat the operation a few days afterwards with the object of still further cleansing the fleece, but it injures the softness of the wool.

Running water is generally preferred with the object stated, but water impregnated with chalk should be avoided, as it causes the wool to be rough and brittle. Some persons, however, prefer a large pond, with clear, stagnant water for washing their sheep, if there is a sound bottom, free from mud—the softness of the water causing it to cleanse impurities from the wool more effectually, and the oily matter, or "yolk," contained in the wool being of a soapy nature, strengthens the wash so much that the greater the number of sheep that are washed in such a pond the better has been the cleansing property of the water.

SHEARING.

It will be found the best plan to shear sheep early in June, if the weather be fine ; for although this operation is often delayed to allow the wool to get heavier, yet there is an advantage found in the growing wool, which prevents the attacks of flies, which in some particular districts are very destructive.

Lambs, at one time, used to be shorn with the sheep in England, and the practice is commonly followed on the continent still, but has been discontinued in this country, the hogget wool being found to be of superior quality to that in those instances where the fleece had been previously clipped.

The number of sheep which can be shorn by one man in the course of a day varies according to the size and the description of fleece, but about fifty head of Southdowns would be the average, and a couple of score of the larger and heavier breeds would be about as many as a good man could shear.

It will be found a good plan to clip off the coarest wool on the thighs and hock about a month before shearing, as this keeps them clean and cool in hot weather. The clippings should be put aside and washed, when uses can be found for them. In shearing, the art is to clip quite close without cutting the skin, as not only is more wool obtained, but it feels softer towards the bottom of the pile. When the skin is cut in the operation the flies immediately settle on the wound for the purpose of depositing their larva, which will ultimately infest the whole carcase with maggots. This is to be prevented by smearing the cut with turpentine or healing salve.

Shearing is generally done under cover, such as upon the clean floor of a barn, but may be equally well performed in the open air. The sheep intended to be shorn on a certain day are penned close to the shearer, who takes each between his legs, placing the animal upon its rump, with its back against him, holding it with his left hand, with his right hand holding a pair of sharp spring shears, blunted at the points, which, being without handles he can manage with one hand, and clips the wool from the neck and shoulders. He then turns the sheep upon its side, and kneeling on one knee holds it down by the pressure of his leg upon its neck, cutting the wool circularly round the body, by which the operation is closely and uniformly executed. The entire fleece is thus cut off at once, and rolled up firmly by another person, with the outside folded inwards from tail to shoulder, and tied together.

The shearing is sometimes done *along* instead of across the body, when the shears moving in a level course cuts the wool closer, and in a more even manner. In the method we have described the shears plying in a curved direction, although it makes the process of shearing more difficult, as well as occasioning a greater risk of nicking the skin, is yet, however, preferred, as giving a neater appearance to the fleece, and is the mode most generally practised.

For shearing sheep it is very necessary to have an able and expert workman who thoroughly understands his business. It is highly necessary that he should be a careful man, for if he cuts the skin the flies will so torment the animal that it cannot thrive. Cuts must, as before stated, be smeared with a little turpentine or any healing kind of salve to guard against the evil effects of these accidents.

The flock should be repeatedly and carefully examined after shearing, in order to find whether any flies have deposited their eggs, and, if so, the tumours should be opened and rubbed with mercurial ointment, the smallest portion of which destroys the insect.

LICE AND TICKS.

Sheep are commonly infested with lice and ticks, and a starling or rook may often be seen on a sheep's back rendering him the good office of picking them out. In order to do away with the necessity of the starlings "shepherding," a mixture should be made consisting of 4gals. of train oil, ½gal. of oil of tar, 1 pint of oil of turpentine. This should be rubbed on every part of the animal to kill the lice and ticks. Tobacco juice will be found equally efficacious. A clean sweep can be made of these pests both in the lambs and sheep by dissolving a pound of arsenic in boiling soapsuds, and then poured into a tub with a large quantity of warm water. This will do for about twenty lambs, in which they should be dipped, and the water afterwards squeezed out of the wool by hand. One dipping destroys the lice, and keeps away the fly and maggots during the summer. It is almost unnecessary to say that care must be taken not to allow the lambs accidently to swallow any of the water by careless dipping. There is no shelter for these vermin on the sheep which have been shorn, and the entire herd will be secured from their annoyance during the whole succeeding year.

MARKING.

Where flocks are apt to mingle, and for the purpose of identification, it is usual to form some distinguishing mark either on the body or head. This is generally done with ruddle (red ochre) in preference to using black marking material, as tar, &c., on the fleece, but the matter being an unimportant one, may be well left to the taste or fancy of each separate individual, except in the usual custom of marking lambs on the off and near sides for wethers and ewes respectively.

MARKETS FOR WOOL.

The markets for wool are very abundant all over the kingdom,
· and there are few agricultural counties without a wool market
being held in one of its principal towns, and there is never any
difficulty in disposing of any amount of produce in this form.

A great uniformity of price now also prevails for wool throughout the kingdom, on account of the rapid inter-communication
which exists through the agency of railways, so that we may say,
for example, that if the wool trade, as respects raw wool or even
yarns, may be flat in Leicester, a brisk trade amongst the cloth
manufacturers of Yorkshire would cause the staple to be actively
in demand at Bradford and Leeds, and the wool would speedily
find its way from one town to another, according to the demand.

The depression occasionally experienced in local wool markets,
so far as consumption immediately on the spot is concerned, now
has but very little effect upon the interests of the grower.

The establishment of the colonial wool sales in Coleman-street,
London, of late years has had a very beneficial influence upon the
English wool trade, though it may seem odd that this should be
the case to an inexperienced person. The wool being of a finer
and better description than most English wool, provokes a great
deal of competition for the best lots, and these sales have been
found to give a firm tone to the wool market, in which all kinds
have participated, when perhaps from temporary dullness of
trade prices would otherwise have languished and gone down.

SALVING.

In the Highlands of Scotland it is thought necessary to smear
the sheep with oil, tar, butter, and other ingredients, for the
purpose of protecting the animals from cold and the scab, for
destroying vermin which may have lodged in the fleece, and
improving the growth of the wool; and although it stains the
latter, so as to render it unfit for receiving very bright or
delicate colours in the course of manufacturing, as the fleece

is not easily cleansed, what is thus lost in price is more than compensated for by the additions gained in its weight.

The opponents of the practice contend that its only advantage is to destroy the vermin, though they admit that in the case of old ewes it prevents the wool from getting coarse and hairy. They consider it injurious to open the wool of the animal at the commencement of the winter season, when it needs all the warmth of its wool, the application of the salve also causing chill, from which it takes ten days or a fortnight to recover.

Its advocates, while admitting that this effect may be produced, argue that the method assists in guarding the sheep against the extreme severity of the weather; and in those instances where some of their stock were left unsmeared at the time of shearing, at the next season the cover of wool was not so good on them.

The operation being both disagreeable and tedious, is often done very negligently, the salve being applied too bountifully in some places, while others are neglected, by which not only is there great waste, but the vermin escape, and the scab increases on those spots which are left untouched.

Four gallons of Virginian tar, with 35lb. of butter, is the quantity usually employed for every forty-five sheep of a flock of Cheviots, and in very mountainous districts that are much exposed, 40lb. of butter and five gallons of tar are often used to fifty or sixty of the small black-faced breed, and if carefully applied the quantity would be sufficient for seventy or eighty sheep.

In applying it, the animal is placed upon a stool of a size sufficient to contain its body either when laid across, or at full length, with a narrow projection at one end, forming a kind of seat, on which the operator sits astride. The wool is then parted longitudinally into rows, without a single pile being allowed to lie across them, so that the salve can be applied directly to the skin. The salve, which should be of such a consistence that it can be taken up by the finger, and yet when drawn along the skin, can be easily rubbed off. The rows thus formed should be at equal distances from one another, in order that the salve in the one may reach through the bottom of the pile to the salve in the next, so that the skin is completely covered, and every individual pile anointed.

No portion of the salve should, however, appear on the wool except at its roots, every other part being quite loose and clean ; when thus carefully laid on, a thin covering will answer every useful purpose. One man, who is accustomed to use this salve, can operate upon a score or five-and-twenty sheep in a day.

When too heavy a coat of salve is applied to ewes, beside the waste caused, there is the further disadvantage that in running out it causes the loose locks of wool which hang around the udder to form themselves into tassels, about the size of a teat. The lamb, in its first endeavours to suck, frequently lays hold of this instead of the teat, and being disgusted with it, does not continue its attempts to suck. To prevent this, the tags of wool should be cut off at the time of smearing.

Some flock-masters have been in the habit of using Gallipoli oil alone, and have found that the wool grew more freely, and was softer when handled than the wool of sheep which had not been smeared ; in the words of one who followed the practice, it is described "as taking possession of the fleece, and keeping the animal in wet weather as dry as a duck; it instantly kills all vermin, and produces a heavy fleece of beautiful soft wool."

The application of tar in exposed situations, guards in some degree against the inclemency of the weather, and also partially prevents the scab. The mountain breeders fear to use oil alone, considering that an oiled sheep rubs the infection of scab into his wool and skin, and so gets the scab confirmed ; while one smeared in the common way escapes it. The tar in its operation appears to retain the oily ingredients within it.

BREEDING.

Ewes are generally fit to breed at fifteen to eighteen months; supposing them to have been dropped in January and tupped at the beginning of the following May, their lambs would be dropped about the end of September, which is only a practice followed for the production of house lamb with the Dorsetshire breed, and eighteen months is the most usual time, and when it is not desirable that the lambs fall until late in the spring, they are not tupped till October. The period of gestation of the ewe varies from 146 to 161 days; in round numbers, twenty-one weeks is the usually estimated time.

During the lambing time everything will depend upon a good shepherd, who, in the case of large numbers, should have assistance, and not be nearly worked to death; for many otherwise good shepherds have been known to neglect their lambs from sheer physical exhaustion, when warmth and shelter were immediately wanted to be furnished to the young animals. The shepherd should be provided with simple remedies only, and the operations of nature left to themselves as much as possible. In cases of exhaustion, gruel mixed with warm ale or gin is useful; and in cases of inflammation, two or three drops of aconite in half a pint of water, repeated every three hours, is a good remedy. The ewe should always have ready access to water. At this season the supply of roots should be curtailed and plenty of dry food given.

For the first month, as the lamb depends on its mother for support, good food ought to be supplied to her to insure a flow of healthy milk. At the end of a month the lamb will begin to forage for itself. Great attention should be paid to see that the lamb thrives, and although when some few weeks old they may lie out at nights during fair weather, when it is rough some kind of shelter should be provided for them; and as soon as they can run they should be allowed to pass through lamb hurdles and pick at the food that is being consumed. The little animal soon learns to feed, and should be made accustomed to artificials, such as bran, dust oilcake, nut meal, and bruised oats.

Early weaning is best for both lamb and ewe, and may take place when they are upon winter vetches.

Tup lambs are generally castrated, if strong and healthy, when about ten days old; and the weather should be dry and mild. Towards morning is the better time for this operation.

Improvement of Breeds.—The qualities which every breeder

seeks to develop in his flock as much as possible are robustness of constitution, rapid and large growth of fleece and carcase, symmetry of form, fecundity, and aptness to fatten. The first improvers of any breed of animals are obliged to couple those which are near akin, in order to develop those salient points which are required, and which it is desired to impart to a whole flock; but when this has been accomplished it is unfavourable to the health and vigour of a flock to breed too much in-and-in, which ultimately will prove injurious to the size and constitution of the stock; and, in order to avoid this result, successful breeders have got some neighbour to allow them to take out fifty of his best ewes and put his best ram with them, from the produce of which they select ram lambs, by which plan, and drafting thirty or forty refuse ewes every year, they are enabled to get up a healthy flock, not too nearly related in point of consanguinity.

The practice of an eminent Scotch breeder, who reared some of the finest Cheviot sheep, is well worthy of imitation. He made a careful selection from the best of his own stock, and bought, without heeding the price, a few from other celebrated flocks of the same breed. He kept these apart, under his own immediate superintendence, so as to study the tendency of each family towards deficiency or excess in different points; and, by judicious crossing between the different families, he succeeded in producing stock which excelled in quality that of both sire and dam. The result of this course was described to have wrought like a charm, in a few years lifting up this individual's stock to an entirely new position, both at the tup shows and to the more remunerative end—in the markets where stock and wool are usually sold.

It is considered that the most vigorous offspring is secured by using shearling rams, and restricting them to about thirty ewes each, and this is a more preferable course to that when older males are put to twice that number of females.

As we have pointed out elsewhere, though the character of both sire and dam can be traced in different degrees in their offspring, that of the male generally predominates. This is a generally well-recognised fact amongst breeders, but it has been the means of leading to an error by employing rams of a much larger size than the ewes they tupped, the progeny being generally of an imperfect form. The reverse should be the case, and the ewe proportionately larger than the ram.

Although, in the case of some of the small breeds of cows, which have been crossed by a shorthorn, they produce calves which ultimately attain a very large size, the principle of the improvement in the stock of sheep resulting from coupling a large ewe with a ram has been thus described by Mr. Clive, who was a successful breeder :

"The proper method of improving the form of animals consists in selecting a well-formed female proportionably larger than the male. The power of the female to supply her offspring with nourishment is in proportion to her size and to the power of her nourishing herself from the excellence of her constitution. The size of the fœtus is generally in proportion to that of the male parents ; and therefore, when the female is disproportion-ately small, the quantity of nourishment is deficient, and her offspring has all the disproportions of a starveling. But when the female, from her size and good constitution, is more than adequate to the nourishment of a fœtus of a smaller male than herself, the growth must be proportionally greater. The larger female has also a greater quantity of milk, and her offspring is more abundantly supplied with nourishment after birth. To produce the most perfect formed animal, abundant nourishment is necessary from the earliest period of its existence until its growth is complete."

The external form of domestic animals has been much studied, and the proportions are well ascertained, but the external form is an indication only of internal structure. The principles of im-provement must, therefore, be founded upon a knowledge of the structure and use of the internal parts. The lungs are of the first importance. It is on their size and soundness that the strength and health of the animals principally depend, for the power of converting food into nourishment is in proportion to their size.

In choosing animals to breed from, a deep and wide chest,

a wide loin and hips, joined with ribs springing gradually wider until they approach nearer to the hip, are of importance. The shoulder-blades should project gradually wider until they approach far towards the hind quarters, which gives more service in the chine, and this is considered a good point in animals intendéd for slaughter, as it enables them to accumulate more meat on that part where it is of the most value.

Disappointment has frequently resulted in breeding from the overfeeding of both sexes, but particularly of rams. The animals have sometimes been allowed to accumulate fat to such an un-natural excess that they frequently fail to propagate their kind, and if they do breed, the lambs are puny at birth, and afterwards badly nursed.

At the same time, during pregnancy and while nursing, ewes ought to be fed liberally, so as to be able to bring up their lambs well and yield good fleeces.

As soon as the lambs are weaned it is a good plan to turn the ewes into poor pastures until about a fortnight before the rams are again admitted to them, when they should be well fed upon succulent food, so that they may be in a rapidly improving con-dition when conception takes place; this course of treatment invariably insuring a satisfactory production of healthy lambs. Although sheep of one breed or another will thrive on all soils, yet it must be remembered that the size and constitution of animals must be adapted to the soil and food of the district they inhabit. Wherever food is produced abundantly of a nutritive description the animals will attain a large frame, but when the produce is scanty in mountainous regions the animals are small in size, and although it may be proper to improve the form and quality of the meat and wool of any aboriginal race, to enlarge the size too much would be very often injudicious, and judgment needs to be practised in these matters.

Flock for Breeding Purposes.—A flock of sheep for the purpose of breeding fat stock should be kept always young by getting rid of the ewes which have produced three or four lambs. Being sound in the mouth and vigorous, they will then fetch a good price, and the loss which attends the sale of old broken-mouthed ewes will be avoided. Of course an exception must be made in the case of valuable animals kept for breeding. All those which have wintered badly and have not done their lambs well should be at once removed.

YEANING.

The ewes when put to the ram, as we have before indicated, should be in fair, but not too high, condition; nor should they be allowed to get fat during the period of their gestation, which may occasion a difficulty in lambing. But, on the other hand, if they are poor in flesh, and weak, they should be put upon better food, as the yeaning approaches, both to give them strength to go through with it, and to get up an abundant supply of milk for the support of their lambs.

This, however, must be done with caution, and of the two it is the safer practice to put both the ewe and lamb on better food after lambing has taken place than before.

The ewes are commonly left to lamb in the open field, without any assistance, beyond the occasional attention of the shepherd, though some careful persons use small huts, mounted upon four wheels, which may be drawn to the flock for the use of the shepherd whenever he may require it; but to afford a little shelter some movable covered pens may be easily formed with hurdles, which can be placed in a warm paddock or anywhere under superintendence near the house on a small farm where no regular shepherd is kept.

The pens should be littered with straw, fern, or leaves, in order to make them perfectly dry, but they should be open to the ewes, so as to allow them to range the field and return at pleasure, and this will be found a better plan than folding. Wherever the yeaning takes place, it should always be effected on a piece of smooth pasture, free from hobs, ruts, ditches, or hollows of any kind.

The most constant and sedulous attention is necessary at yeaning time, for occasionally, in very severe weather, when it has been long in its completion, the ewe is so exhausted as to

need some warm gruel, and to be housed from the rest of the
flock until she is sufficiently recovered to pass muster with the
others.

WEANING.

The time of weaning has to depend upon that in which the
lambs have been dropped, for if early in the season, and the
weather continues cold, they are mostly allowed to remain with
their mother till it becomes warmer.

The early part of July is the common time, and it should
never be deferred beyond the end of that month, unless the
lambs have been dropped late in the spring. Nothing more is
required to be done than to drive the lambs so far away from
the ewes that their reciprocal bleatings may not be heard one
by the other.

The weakly lambs should, however, be separated from the
stronger ones, and if possible should be placed in a different
pasture, and where breeding is carried on, close supervision
should be exercised in the selection of ewe lambs, and any
that may be ill-formed, or defective in any points, should be
sold and cleared off, or they will assuredly deteriorate the quality
of the flock.

Again, in sorting out the lambs the wethers and ewes should
be differently distinguished by being marked on the off and near
side respectively. As before mentioned, marking is done either
by the ruddle, in the fleece, or with tar, or by a brand upon the
cheek and notches in the ears. In the latter case, marks cannot
well be obliterated, and where there is any chance of identification being difficult the latter method is sometimes resorted to.
At the same time, although a notch cannot be obliterated, another
could be added by any dishonest persons wishing to tamper with
marked sheep. After the lambs have been taken away, the ewes
should be milked three or four times, at longer intervals from the
first commencement of milking, so as to dry them up by degrees
with as little inconvenience to them as possible ; or, if this is not
done, they should be confined for three or four nights in a fold,
and in the daytime be permitted to range over a poor pasture.

The bare keep assists in preventing injury from an accumulation of milk, but attention should be paid to the condition of their udders, and those which appear full of milk should be partially drawn.

It should be remembered that where this is neglected on account of the trouble that it entails, it both occasions considerable pain to the animal, and is sometimes attended with serious consequences. At one time it was a very common practice to milk ewes for the purpose of making cheese, but it is now seldom done except in very remote districts where old methods and old customs are handed down from one generation to another.

The lambs, when taken away from the ewes, should be put upon the best pastures on the farm, or upon the second crops, or aftermath, of good clover and saintfoin, spring tares, or any abundant, succulent, nourishing feed that will promote their growth, which at this period is a very important point, and the expense should be scarcely considered, as it is very desirable to push them along without receiving a check. If their progress is retarded, it cannot very easily be recovered.

The ground should be occasionally shifted, so that they are kept upon nice fresh feed, and it pays to watch them carefully until the herbage begins to fail, when they should be put upon rape, turnips, &c., for the winter.

AGE OF SHEEP.

Sheep become "broken-mouthed" in their sixth or seventh year, their natural age being generally nine or ten years. When deprived of the powers of easy mastication they fall off in flesh, and, therefore, do not pay to keep. During the first year of their lives their teeth are of small size, but at fourteen or sixteen months old they renew the first two, and two more every year until the fourth shearing, at which time they become full-mouthed. Those animals which are well fed renew their teeth the earliest. The age of sheep is usually counted from their first shearing.

F

Prime Mutton.

Mutton is really not in its prime till six years old for hanging and eating; and it is quite a mistake to speak of young mutton as being desirable, for then, properly speaking, it is neither lamb nor full-flavoured mutton; but the system of turnip husbandry, which has sprung up of late years, has forced sheep into earlier maturity. An exception even now prevails, however, in killing heath-bred Cheviot and black-faced mountain sheep, which are not reckoned fit for the butcher until the wethers attain three years of age, rising four, and the ewes five, rising six.

Designation of Sheep at Various Ages.

The male sheep is called a ram or tup before he is castrated, after which he becomes a wedder, or wether; the latter almost universally employed by the butcher, and the female an ewe, both being called lambs until they are weaned, after which the males are termed wedder-hogs, hoggets, hoggerells, or tegs, from the time of weaning till the first shearing; and the females ewe-hogs, or gimmers. After the first shearing the males are called shearling-hogs, or Dinmonts, and the females gimmer-shearlings, or theaves, after which they are known as two shear, three shear, or four shear wedders or ewes, until they are full mouthed; when the ewes, if drafted out of the flock, either as being barren, or to be fatted, are called cast or yelled, and, when turned six years old, are called crones.

CHAPTER V.

SHEEP (Continued).

Diseases and their Treatment—Signs of Health in Sheep—Blindness—The Blood—Inflammation of the Bowels—Bronchitis—Calculi in the Bladder—Catarrh—Consumption—Dysentery, or Braxy—The Epidemic—Diseases from Exposure to Wet—Flies—Flux, or Scouring—Foot Rot—Hoving, Hoove, or Blasting—Obstructions in the Gullet—Pining—Pleurisy—Pneumonia, or Inflammation of the Lungs—Rabies, or Canine Madness—Redwater—The Rot—The Scab—Sheep Pox—Slipping the Lamb and Protrusion of the Uterus—Sore Teats—Sore Udders—Sturdy, Dunt, Staggers, &c.—Ticks or Sheep Lice—Diseases of the Urinary Organs—Worms.

Diseases and their Treatment.

There may be a few other disorders and ailments, but those we describe are the "chief ills to which (sheep's) flesh is heir to," and, as will be seen in the following, most of them are to be warded off by preventive treatment, if resorted to in good time. Watchful care and attention with sheep will always be rewarded, and if there are signs of interrupted health—however slight they may be—a careful examination of the animal, or animals, should be made at once.

A watchful shepherd's or master's eye will detect any failing of health at once, and it has been correctly observed by Price, on sheep management, that "when he is in good order he carries his head high, the eye is of clear azure, with a quick and lively aspect; the mouth is clean and of a bright red, the gums ruddy, the teeth fast, and the muzzle dry; the nostrils damp without being mucous, the breath free from any bad smell, the feet cool, and the dung substantial. The hams are strong

F 2

and the limbs nimble; the wool firmly adhering to the skin, which ought to bear a reddish tint, with a soft mellow feel of suppleness, and, more especially, the appetite should be good."

Blindness.—Cold and wet, producing inflammation, is sometimes the occasions of blindness in sheep. The only treatment which can well be followed with any chance of success is to bleed under the inner angle of the eye, on the side of the nose. The method of doing this is to lay the sheep on its back, and have it held firmly in position, while the operator, with a sharp instrument, cuts the blood-vessels on the inner angle of the eye —one or both—at about $\frac{1}{4}$in. below the angle.

The wound must be deep enough to allow the blood to flow freely, which if it does not do upon the first incision, another cut must be made a little lower or a little higher. Epsom salts also should be given to open the bowels freely.

The eye-balls of the sheep will be found to present a singular appearance, the whole surface being of a light blue colour.

Blood, The.—Sheep fed upon very rich pastures are subject to this disorder, which is sometimes called the "yellows," from the flesh turning yellow after death, and frequently arises from sheep eating red clover while it is in blossom.

The sheep attacked may be seen to separate themselves from the rest of the flock, as if in pain, and stretch out their fore legs to find relief. The breath is short, their eyes appear heavy, and the abdomen seems convulsed. The disorder, being of an inflammatory nature, is often suddenly fatal, and demands similar remedies to that prescribed for braxy.

Bowels, Inflammation of the.—Inflammation of the bowels is rare with sheep, except in the form of braxy, but when it does make its appearance the symptoms very much resemble those of influenza in the first instance, the eyes being inflamed, closed, and running with water, breathing laborious, with constipated bowels, loss of appetite, and staggering gait. The liver gets decomposed, and the spleen enlarged and gorged with blood.

Bleeding should be resorted to in the early stage, followed by aperient and febrifuge medicine.

Bronchitis.—Sheep are not so often affected with this disorder as cattle, but the treatment should be the same, *i.e.*, bleeding, a seton inserted in the brisket, and the administration of aperient and febrifuge medicine of a mild character. Half-a-pint of lime water for a sheep, and quarter of a pint for a lamb, should be given morning and evening ; or a couple of teaspoonfuls of salt dissolved in water, the treatment to be continued for a week.

Calculi in the Bladder.—Sometimes small stones are found in the ureters after death, but calculi in the bladder are seldom found, the small stones being discovered most frequently in the urethra, where they form an obstruction to the passage of the urine and occasion great pain, which results in inflammation, and death follows unless relief is afforded.

Where a calculus is suspected, it is advisable to mingle acids in the food with the view of dissolving it, as it consists almost entirely of phosphate of lime. Linseed tea made slightly acid with sulphuric acid might be given and repeated for some time, but if the disease has reached an advanced stage, an operation will be necessary to remove the calculus, and, after it, soothing and aperient medicines should be given.

Catarrh.—At the fall of the year, particularly in a wet season, sheep are very liable to catarrh, the seeds of which are sown by their being exposed too much to the weather after shearing, before the fleece has sufficiently grown to afford the necessary protection.

The membrane lining the air passages becomes inflamed, and an increased discharge takes place, which brings on a cough. Sometimes the disorder will last several weeks, and then get gradually well, the constitution of the sheep having thrown off the ailment, but occasionally inflammation will extend to the lungs, and turn out fatally. Plenty of shelter and good nursing are wanted in these cases, which, when mild, a little gruel will often cure ; but if the symptoms are severe the following should be given :

Epsom salts	½oz.
Ginger	1dr.
Nitre	1dr.
Tartarised antimony	½dr.

dissolved in gruel.

In very severe cases bleeding from the neck must be resorted to.

Consumption makes its first appearance very often with ewes that have lambed. But there is every reason to believe that wet pastures, or exposure to wet in one form or another, first laid the foundation of this disease, which was prevented from being manifested on account of the parturient condition of the ewe. Abcesses are found in cutting into the abdomen, adhering to the bowels and mesentery, as well as in the liver; whilst the latter is smaller and paler than usual. Abcesses are sometimes found in the lungs also. When matters have gone this length there is no dealing with them with any hope of effecting a cure.

Dysentery, or Braxy, being inflammation of the coats of the intestines, is much more serious than diarrhœa. It mainly arises from undigested food remaining in the stomach, which the animal, as it does not immediately lose its appetite, adds to, when fermentation takes place, resulting in inflammation, by which many young sheep are carried off.

It is sometimes known by the name of the *sickness*, and commonly makes its appearance about the end of autumn, and is occasioned, it is thought, by the herbage becoming more dry and astringent, as it ceases when winter sets in, being more prevalent on dry pasture, where there is both heath and fern, than in low lands, a sudden change from a succulent pasturage to a dry one having been often observed to produce it. Exposure to wet and cold after travelling will also bring it on.

Its symptoms are—dung hard and scanty, though frequently discharged, and covered with mucus and blood, accompanied with an offensive smell, dulness, uneasiness, want of appetite, quickened pulse and respiration, with dry nose and skin.

Bleeding, which may be effectually done by cutting the tail, is prescribed, though most veterinary surgeons would, perhaps, bleed in the neck or the fore leg. Glauber salts or castor oil should also be given, and the sheep put upon turnips during the day, and placed in a dry yard or fold during the night.

Another course of treatment is to give the animal affected

linseed gruel, so, as it were, to lubricate the intestines, as well as afford nourishment; and afterwards give the following medicine:

Linseed oil ... ... 2oz.
Powdered opium ... ... 2gr.

Next day the opium may be repeated with a scruple of powdered ginger and two scruples of gentian root, and the oil again be given if required.

Epidemic, The.—The disorder which passes under this name is very similar in appearance to the foot rot, but proceeds from a different cause, arising from fever in the system, the symptoms exhibiting themselves either in the shape of bladders in the mouth or soreness of the feet. The first appearance of the disease is generally, however, in the feet, a number of sheep falling lame at the same time, the mouth of sheep being rarely affected.

The feet get hard previous to the formation of matter, the result of inflammation, and in some instances the disease proves very troublesome. The best treatment in its early stage, while there is fever, is to give Epsom salts, so as to cool the system and shorten the duration of the fever, and afterwards to treat in a similar manner to the method adopted in foot rot.

Exposure to Wet, Diseases from.—In the spring which succeeds a wet autumn, there are frequently complaints heard of ewes casting their lambs six or eight weeks before they are due, the lamb generally being dead.

The abdomen is found distended with water, and the ewe both suffers during labour and afterwards. Puerperal fever follows, and is often fatal. This is more particularly the case when the heavy ewes have been kept upon turnips previous to lambing, but this many farmers make a practice of avoiding. The turnip itself contains about 90 per cent. of water, while in wet seasons there is a good deal of wet hanging to the root and appendages as well. It is best, therefore, where it can be done, to keep the ewes before lambing without turnips, or at least give them a little corn, linseed, or rape cake in addition.

Flies.—Sheep are very much troubled with flies, which lay their eggs upon the skin, and these, when hatched, produce maggots during spring and summer, and, if not carefully looked to, are often attended with disastrous results.

Hogg, the Ettrick shepherd, has described an occasion when he went amongst a flock of sheep upon which the flies had settled in such enormous numbers that when disturbed, and they rose up, he, and the persons with him, could with difficulty see each other; but upon some sheep, which had been anointed with whale oil, being turned in amongst them, to his utter astonishment, in less than a minute, not a fly was to be seen!

Hogg also was of opinion that either a certain habit of body, or some kinds of food, at times, give the excrement and perspiration of an infected sheep a peculiarly rancid and loathsome smell, so that when near them it is not difficult to discover those that will be soon infected. The most healthy sheep are often attacked by this disposition of body, which is frequently accompanied by diarrhœa.

There are several kinds of flies which infest sheep during the summer months. The small common flies are very troublesome in enclosed and woody parts of the country—more especially in warm and showery weather. These find out any luckless sheep that has a scratch or sore upon its skin, which they attack in warms and never leave till they are gorged.

At these seasons sheep should be driven to the uplands if possible, but if not, it is a good preventive to wash them after being sheared, and once or twice more during the summer with soap-suds, liberally impregnated with asafœtida, empyreumatic oil, or similar applications which will retain a disagreeable scent.

The most offensive flies are the sheep-maggot flies, or large flesh flies, or blue bottles, of which there are several kinds, black, white, and those of a greenish colour, which bear a strong family likeness to one another, but which come to maturity at different periods. These are produced by putrid carrion in the first place, and are attracted as soon as they can fly by any strong odour like that of the perspiration of sheep in hot and misty weather, or in morning dew.

The fly deposits its eggs upon any likely spot to develop them

into life—near the root of the tail, around the anus, or where
excrement may have been hanging in these parts, specially
attracts them, though the back is sometimes visited. It is, therefore, desirable to clip the rump and buttocks close previous to the
ordinary shearing time, as well as a couple of months afterwards,
and wash with the liquid prescribed below, which is likely to
prevent breeding of maggot without hurting the wool:

Arsenic, finely powdered	... ...	1lb.
Potash	... ...	12oz.
Common yellow soap	... ...	6oz.
Soft water	... ...	30 gallons.

These ingredients must be boiled together for a quarter of an
hour, the person who does it taking care to inhale as little of the
steam as possible.

Many experienced shepherds have used the following with
good effect : An ounce of sublimate, to which is added three
tablespoonfuls of turpentine, dissolved in a gallon of water.

The sheep are subject to the attacks of flies from May till September ; and unless a sheep which has become infected is immediately attended to, within two or three days it will be perforated
by some thousands of maggots, and if neglected perhaps within a
week become a lifeless mass covered with these larvæ.

Serious mischief can, however, be prevented by constant attention in clipping, cleaning, and anointing the maggoty parts.
Train oil with sulphur is a useful application, and pepper will
destroy them, and where oil of turpentine can be used, it will
be always found a capital remedy.

Sheep hang their heads and betray great uneasiness when
struck by the fly. They stamp violently with their feet, move
their tails in a peculiar manner, and draw up their bodies in a
way which indicates their restlessness, even before the skin is
wounded, which will serve as a sufficient proof to a watchful
shepherd when they have been struck ; and a sheep displaying
these signs of inquietude should be at once caught, and examined.

This should be done closely, and if only the nits have been
formed, they can be soon cleared away, and the parts affected well
washed with soap suds and urine, or goulard water. If, however,
maggots have made their appearance, the wool surrounding any

damaged part of the skin should be cut away, and the part well washed with the liquor mentioned. It is best to avoid cutting the wool away if it can be done without, as it gives an unsightly appearance to the sheep, and also lays bare a vulnerable place to the attack of small flies. Should the maggots have perforated the skin and have got under it, the point of a penknife should be used, so as to remove them, and the wound healed by any common ointment fit for the purpose; or, in bad cases, a good ointment is composed of the following :

Pure quicksilver	½lb.
Venice turpentine	½lb.
Arsenic	½oz.
Neat's-foot oil	½ pint.
Hog's lard	1lb.

mixed thoroughly together by well rubbing in a mortar.

Flux, or Scouring.—Diarrhœa is frequently caused by cold and wet, but it often occurs with sheep of weakly constitutions which are put first upon poor watery food and then removed to rich pastures. If arising from wet, the animals should be removed to a dry situation and be supplied with sound hay, or good dry food of the kinds previously recommended. If occasioned by being put on to rich land, they should be changed to short grass and a little corn given to them. Chalk should also be put into the water they drink.

If the purging becomes violent, a drachm of rhubarb in half a pint of warmed milk will be found useful. If this treatment does not succeed in stopping it, the dose should be repeated, with the addition of a few drops of laudanum.

The following is also considered a good standing medicine to be resorted to :

Catechu (powdered)	4dr.
Prepared chalk (powdered)	1oz.
Ginger (powdered)	2dr.
Opium (powdered)	2dr.

to be mixed with half a pint of peppermint water, and given twice a day. Two or three tablespoonfuls is a dose for a sheep, and half that quantity for a lamb.

Lambs, indeed, are more subject to diarrhœa than sheep; in the first place, commonly owing to the change of food after weaning, and the functions of the stomach being more severely taxed, the new food being too stimulating for their digestive organs.

Amongst lambs, the disorder is sometimes termed *gall*, at the season of the year when the grass is springing, which they eat after a few warm showers, sometimes with fatal effect. If not old enough to eat hay, a little pea-meal, or barley-meal, and some astringent cordial drink can be given with advantage. There are two distinct *presentations* with lambs, called *green skit*, from the colour of the fæces, arising from the change of food mentioned; and the *white skit*, which is of a different nature altogether, and arises from constipation rather than looseness. This owes its name to the pale colour of the fæces, and is due to coagulation of the milk in the fourth stomach, where it accumulates often to the weight of several pounds, while the whey passes off by the bowels, and gives, where the right cause is not attributed, a deceptive appearance to the dung.

The symptoms of this latter are, in addition to the pale colour of the fæces, heaving of the flanks, dulness, hardness, and distention of the abdomen, and sometimes costiveness.

The proper treatment is the administration of alkalies to dissolve the hardened mass. Half an ounce of magnesia, dissolved in water, or a quarter of an ounce of hartshorn in water, or both these combined, and administered in proportionately smaller quantities, should be given and repeated, followed by Epsom salts. After this the cordial recommended above should be given, in a somewhat larger quantity of water.

Another phase of looseness of the bowels in lambs is when the dung is of a very dark colour and tinged with blood, caused by disease of the mucous coat of the intestines, which, after death, is found of a very dense hue, and nearly black.

As this disease attacks the lamb when it has partaken of no other food than the ewe's milk, it is communicated through this medium, though the ewe herself gives no signs of there being anything amiss with her. When it is discovered the food of the mother should be changed as quickly as possible, and more dry food given to her, a little oilcake and salt proving a useful addition to it.

The progress of the disease is, however, so rapid that a lamb has been found dead in the morning which gave no signs of illness on the day previous ; and in the discovery and treatment of such cases is afforded the best proof of a watchful shepherd's care.

Foot-rot. — This disease, which is partly contagious, and is supposed to be communicated by sound sheep treading upon matter discharged from those having diseased feet, is seated in the vascular parts which connect the hoof with the bones of the foot, and generally arises from exposure of the hoof to too great an amount of moisture, and is more prevalent in low marshy districts than in uplands. Chalky soils, or ground that has been strongly manured with lime, are thought to peculiarly exempt from attacks the sheep placed upon them, and it has been considered a good plan to drive those sheep which are supposed liable to infection into a pen strewed with quicklime as a precaution against the infection.

The foot of the sheep being naturally adapted to a dry soil the wear of its lower surface is equal to the growth above, but when it is constantly immersed in wet it grows soft and becomes elongated, and is then easily penetrated by gravel or small stones, and the upper part where it is thinnest becomes detached from its connections. Soreness and inflammation are thus produced, with subsequent suppuration, and sometimes ulceration, which, running under the horn, occasions the hoof to be sometimes entirely lost ; so that the animal crawls on its knees in search of food, and by the exertion, combined with pain, becomes so exhausted that it perishes if not speedily succoured.

A slight halt is the first indication of the disease, and that shortly increases to positive lameness, the foot being hot, with an appearance of swelling about the heel, or between the toes, and the hoof is in a slight degree loose. The animal affected should be at once removed to an open shed or pen, with some clean straw litter, and kept separate from the sound sheep, and the foot well washed with soap and water.

The foot should be minutely examined, after the dirt has been removed, so as to find out the injured part, [which, although not always apparent to the eye, may be detected upon pressure with the hand. The hoof must be carefully pared away round the affected part with a sharp knife, using precaution not to cut any sensible part of the foot, and anointed with one of the ointments hereinafter mentioned, which should be laid on with a feather in every crack or crevice that may be affected, and the animal confined until the application becomes dry, but no bandages should be used, as the disorder is of an inflammatory nature.

Previous to any subsequent application, the foot should be kept clean and washed with lime water, the dressings to be repeated within two or three days.

The following are all considered good recipes for foot-rot :

Tar	8oz.
Lard	4oz.
Oil of turpentine	½oz.
Sulphuric acid	½oz.

The two first ingredients should be melted together separately, and the two remaining ones added slowly, and carefully, afterwards.

The following composition is recommended as a speedy one in its operation :

Turpentine	2oz.
Diluted vitriol	½oz.

to be stirred up before using.

Where an active application is sought for, the annexed is held in high estimation :

Corrosive sublimate	1oz.
Blue vitriol...	2oz.
Verdigris	2oz.
White copperas...	½oz.

dissolved in half a bottle of white wine vinegar, and applied with a feather.

Another :

Verdigris	1oz.
Blue vitriol...	0oz.
Spirit of turpentine	1 gill.
Distilled vinegar	3 gills.

In damp weather a mixture of tar and salt has been used with success.

The following has been used very successfully in Scotland:

Corrosive sublimate	... ...	½oz.
Sulphate of copper	... ...	2oz.
Verdigris	... ...	1½oz.
Alum	... ...	2oz.
Sulphate of zinc	... ...	½oz.
Muriatic acid	... ...	2oz.
Charcoal	... ...	½oz.

pounded as small as possible in a mortar, and mixed with half a bottle of distilled vinegar.

Some flock-masters, when the disease has made its appearance, in order to protect the animals from undue moisture, cause their sheep to walk every day over a dry and smooth surface, upon which lime has been strewn.

A somewhat similar treatment to that prescribed for foot-rot should be resorted to when soreness arises from travelling and there is irritation of the biflex canal between the claws.

Hoving, Hoove, or Blasting.—Sheep, in common with cattle, are liable to this distension of the rumen by gas, caused by the fermentation of the food which they eat upon too sudden a transition from poor pasture into succulent artificial grasses, as red or broad clover, &c.

This result is best prevented by giving them a little dry food in the morning of the first three or four days, so as to partly fill their stomachs. The disorder calls for prompt treatment, and the hollow probang should be passed into the rumen, so as to allow the gases to escape through it. Half a pint of linseed oil given to each sheep with a horn occasions them to vomit, and is said to have never been known to fail.

If there is nothing else at hand, some salt, about as much as would be contained in a dessert spoon, may be dissolved and poured down the throat; or a drachm or more of chloride of lime dissolved in water. Sulphuric ether in doses of two drachms is also an effectual remedy. When the case is pressing, and there is no time to administer medicine, it is necessary to plunge a trochar or a pen knife into the rumen through the flank. If the latter, a small tube formed of the stick of elder or a quill should be inserted to allow the gas to escape.

After treatment is often essential, as indigestion is prone to succeed, and it will be found necessary to administer the following draught :

Sulphate of magnesia	2oz.
Gentian	1dr.
Ginger	2dr.
Chloride of lime	1scr.

dissolved in warm water or gruel.

Caution should afterwards be practised with respect to the diet the animals have to subsist upon, of which sufficient hints have been given previously. Some sheep-farmers consider it good practice to sprinkle salt upon very luxuriant herbage before turning sheep upon it.

Obstructions in the Gullet.—A fragment of turnip imperfectly masticated, or too much taken at one time, will sometimes occasion obstruction in the gullet, but these accidents are of much less frequent occurrence in the sheep than the ox. The symptoms, as may be supposed, are distressed breathing and threatened suffocation, the obstructing body pressing on the windpipe, and thus impeding the passage of air to and from the lungs.

The probang should be passed gently into the gullet over the root of the tongue, the sheep's head being elevated, while firmly held between another person's knees. Care should be taken in using the probang (previously oiled) not to lacerate the side of the esophagus, and in most instances the obstruction can be pushed into the rumen.

The sheep should afterwards be kept without food for a short time, and a dose of linseed oil can be given with advantage. Should, however, the use of the probang fail, and there is danger of the animal's choking, nothing remains but to cut into the esophagus and remove the obstruction ; the operation is necessarily a dangerous one, and the wound both in the gullet and the skin must be carefully sewn up, and for some days afterwards food must be given in a liquid form.

Pining.—This is a disorder of a very peculiar nature, which is not met with in districts where green succulent pastures abound, nor upon those of calcareous or sandstone soils, and is accredited with making its appearance simultaneously with the more extended system of drainage which now prevails, and is attributed to the astringency of the herbage of very dry pastures, the disorder being more prevalent in very dry summers than when the season has been showery. It has been most commonly met with on the pastures of the Cheviot Hills and the mountainous land in the counties of Roxburgh, Selkirk, and Peebles, its appearance being attributed, in some instances, to the destruction of moles, whose labours, although they disfigure a pasture, are the means of producing soft succulent plants upon the ground turned up by them.

Hogg, the Ettrick Shepherd, says the farms most liable to the disease are those which in former years were wholly overrun with moles and which are now intermixed throughout with great ridges and flats of white and flying bents—which last are the bane of the flocks, and that exactly as the laxative and succulent herbage prevails over the dry and benty, the effects of the pining will be less felt. The Ettrick Shepherd thus describes its effects: " On the genuine pining farms sheep do not take it by ones and twos, but a whole flock at once. It is easily distinguished by a practised observer, the first symptoms being lassitude of motion and a heaviness about the pupil of the eye, indicating a species of fever. At the very first the blood is thick and dark of colour, and cannot, by any exertion, be made to spring, and when the animal dies of this distemper, there is apparently scarcely one drop of blood in the carcase. It lives till there does not appear to be a drop remaining, and even the ventricles of the heart become as dry and pale as its skin.

" It is most fatal in a season of drought, and June and September are the most deadly months. If ever a farmer perceives a flock on such a farm having a flushed appearance of more than ordinarily rapid thriving, he is gone. By that day eight days —when he goes out to them again—he will find them all lying, hanging their ears, running at the eyes, and looking at him like so many condemned criminals. As the disease proceeds the hair on the animal's face becomes dry, the wool assumes a bluish cast, and if the shepherd have not the means of changing the pasture, all those affected will fall in the course of a month."

Salt placed in lumps upon the ground, accessible to the sheep, would have a beneficial effect, and laxative medicines are given with advantage, for if the bowels are opened a change appears to take place in the constitution of the animal; but a change of pasture to one of a more succulent nature, as to clover, in order to check the costive habit, is the most effective remedy.

It is said that few of the ewes that have been attacked in autumn with this disease have lambs the next year; but, fortunately, the disorder is not one that is very widely spread, but appears confined to a somewhat narrow area.

Pleurisy.—Although this is inflammation of the membrane which lines the chest, and one often connected with diseases of the lungs, yet it is quite a distinct disorder, being characterised by symptoms of inflammation, pain, and fever, there being greater external warmth than in the previous disease, and the pulse is stronger.

The disease is produced by the system receiving a chill, and frequently follows sheep-washing, some breeds being more subject to it than others, among which may be mentioned the Leicester. Bleeding is generally resorted to, followed by aperient and febrifuge medicines. Some recommend a seton in the brisket.

Pneumonia, or Inflammation of the Lungs.—This is also a somewhat rare disease, but does occasionally happen in low and damp situations. Congestive inflammation is rapid in its progress, being attended by accelerated breathing, and a quick and weak pulse; the lungs, after death, being found congested with blood and black in colour.

When pleurisy has been engendered, the animal shows signs of being in pain, the pulse being strong as well as quick.

Bleeding in the neck is prescribed for these diseases, and this can be better sustained in the latter than in the former case. Purgatives should be given, followed by sedative medicine like the following:

Nitrate of potash ... ... 1dr.
Tartarised antimony ... ... 10gr.
Ipecacuanha ... ... 5gr.

which will be a daily dose for a sheep.

G

Rabies, or Canine Madness.—The bite of a rabid dog will occasion a violent and fatal derangement of the nervous system of a sheep, the symptoms of which exhibit themselves from two days to six weeks after the injury has taken place.

The disease is invariably fatal, and there is no treatment upon which the slightest reliance can be placed, except in the cauterisation of the wound shortly after the injury, either with a hot iron or lunar caustic. In order to examine the subject thoroughly, it becomes necessary to cut off the whole of the wool, and to carefully inspect the whole body, applying the caustic, iron, or knife to the smallest scratch which may have been inflicted by the dog. If the bitten part is carefully cut out there is no danger of the flesh becoming unfit for the butcher.

The symptoms are—propensity to mischief, a disposition to ride other sheep, or by butting furiously ; followed by nervous irritation, quickened respiration, twitching of the muscles, sometimes drowsiness, a great difficulty in swallowing, and a discharge of saliva from the mouth. This saliva is infectious from a rabid sheep, and care must be exercised in handling them.

Redwater.—Redwater in sheep is a different disease to that bearing the same name with cattle, which, in the case of the latter, signifies a discharge of dark-coloured urine.

In sheep it arises from an increased action of the vessels of the peritoneum, or serous membrane, which lines the abdomen internally, and the bowels, &c., externally, which causes an effusion of red serum or water in the abdomen outside the bowels.

Young lambs are subject to it before they are weaned, as well as afterwards, and sheep are occasionally affected. Its victims are mostly attacked in the night, more particularly when feeding off turnips and the ground is covered with a hoar frost. There are some doubts whether the disease is occasioned by lying on the ground or from eating a quantity of cold watery food.

The following medicine will be sufficient for eight or ten sheep, or double that number of lambs :

Sulphate of magnesia	1lb.
Ginger (powdered)	1oz.
Gentian (powdered)	1oz.
Opium (powdered)	½dr.

which should be dissolved in warm water or gruel.

If a flock feeding upon turnips shows signs of being infected with this disease, it will be best to remove them from the turnip field, or only allow them to remain there during a portion of the day ; and in the case of lambs, the progress of the disease is so rapid that it is generally considered best to kill the animal at once.

A good preventive of the disorder in lambs is the following :— two ounces of myrrh boiled in sixty tablespoonfuls of ale, a dose of three tablespoonfuls to be given to each lamb at Michaelmas.

Rot, The.—At one time the rot was the most destructive malady to which sheep were subject, but of late years, until recently, it has not been so prevalent. The disorder (also called cothe) consists of the presence of parasites in the liver, termed "flukes," or "plaice," which are triangular flat-shaped worms, somewhat resembling in shape the fish so called, and vary from an eighth of an inch to a quarter of an inch, or more. They are found floating about in the biliary ducts, apparently feeding upon the bile, and preventing it from performing its allotted functions.

The animalculæ from which these disagreeable invaders proceed are deposited in the herbage of marshy places, whence they are taken up with the food, and are afterwards developed.

Among sheep bred upon wolds and downs the disease is almost unknown, but sheep which have been removed to low ground that has been covered with water during the winter may be infected in a single night. However the case may stand, one thing is perfectly clear, that it is dangerous to turn sheep during the summer months over land which has been flooded.

In the early stage of the disorder the animals appear lethargic, dull, and indifferent about feeding; and when advanced still further the presence of the disease can be felt by handling. The loins feel loose and flacid, the skin has lost its healthy tinge of red, and has assumed instead a pale hue; the eyes wear a dull, sickly aspect, and the animal is more restless than usual.

The first symptoms of rot, however, do not give cause for alarm; on the contrary, the sheep to all appearance for a short space of time look as if improving; but when this stage is passed the unfavourable one commences, which does not immediately affect the flesh, for sheep are continually purchased to all appearance in good health which, when killed, are found with their livers full of flukes, but the meat is quite sound and well flavoured.

Salt has been found the best preventive, and if sheep are obliged to be placed on damp pastures they should be carefully moved to a dry situation before evening, and fed partly with hay. Preventive measure are, indeed, the only ones to resort to, for when the disease is established there remains nothing but to hurry on the fattening process with the most nutritious food, in which linseed and oilcake should largely enter.

Damp marshes, in close proximity to the sea, where the water is salt, do not cause the disease, and this fact would seem to justify the belief that salt is fatal to the growth of the parasites. A good supply of salt, therefore, in such situations where rot might be supposed to generate is, doubtless, beneficial, though a certain amount of caution must be used in the case of heavy ewes, as salt has a tendency to produce abortion, though rock salt placed in the troughs for them to lick could not be attended by any serious consequences.

Scab, The.—The scab is occasioned by an impure state of the blood, chiefly arising from bad feeding in wet ground. It is a cutaneous eruption analogous to the mange in other animals, and generally begins with the itch, the symptoms of which are at first indicated by small white spots on the wool, which may be seen long before the animal betrays any signs of uneasiness or the smallest eruption appears upon the skin. It is highly contagious, and the best remedy is any of the common applications of sulphur.

If a cure is not effected by the sulphur and the intense itching continues, blotches make their appearance upon the skin, which wears a fretted appearance, and, if neglected, soon discharges a fœtid ichor which turns to scab. The animal loses flesh, the wool becomes foul, and, if not cured, the sheep sinks under the continued irritation to which it is subjected.

An infected animal should be immediately separated from the flock, and, if taken at an early stage, the disorder can be removed by any of the common washes formed of tobacco juice, soft soap and urine, or salves of which brimstone forms the chief ingredient. In more virulent cases, the ointment recommended for sheep-maggot may be used.

The disease being infectious, sheep in the highest condition sometimes contract it from coming in contact with posts and fences against which scabby sheep have rubbed themselves, so that it is a necessary precaution in regard to these to wash any cotes or buildings where infected sheep have been previous to the introduction of sound ones.

The disease is by no means incurable, even in its worst stage ; but if it is found that, when destroyed in one part it makes its appearance in another, it is a sure indication that the system or whole mass of blood is infected, and internal remedies become necessary. Purgatives of a cordial nature should be given, but, if the disease arises from poverty, a good supply of wholesome food is the first requisite.

Sheep-pox.—This disorder is the most prevalent during winter, when animals are shut up, and, perhaps, kept very hot. Its symptoms exhibit themselves in the sheep becoming dull and loathing their food. The head, eyes, ears, and gums are swelled ; hard white tumours appear in the groin as well as under the joint of the shoulder, and pimples break out three or four days afterwards in various parts of the body, those on the naked skin between the thighs, where the wool is short, being first affected, until the whole body is covered.

The animal swallows with pain and breathes with much

difficulty at this stage of the disease, and, if it progresses, the pimples become inflamed, enlarge, and burst. The discharge hardens upon the wool, and causes it to form in large lumps, which are rubbed by the sheep, causing the scarfskin to peel off in large pieces full of holes.

When the disorder first makes its appearance, the sheep need to be kept warm, be well littered with straw, and fed upon hay with a little salt, their water being given to them lukewarm. Bleeding is practised by cutting the ear, and the cotes are thoroughly fumigated for five or six days by burning stems of garlic, which occasions a great discharge from the nostrils, upon the effects of which many experienced shepherds repose great reliance.

Whatever be the season, when the distemper abates the wool should be clipped in order to assist the drying of the pustules and favour the growth of a new fleece. It is said that the new wool which comes after the distemper is finer in quality and more silky than upon any previous or subsequent growth.

Sheep-pox so closely resembles scab that it is not known in this country frequently as a separate disease; but in some parts of France, especially in those districts adjoining the Pyrenees, whole flocks have been regularly inoculated for it, the method of doing this being to place the skin of a sheep which has been infected upon the floor of a cote, into which the lambs are driven when about six months old. They voluntarily rub and roll themselves on the skin, and symptoms of the disorder soon make their appearance, from which they speedily recover, much in the same manner as that of the human subject who has been inoculated with lymph from the cow for small-pox. The sheep are never afterwards attacked; and where this plan is practised, it very rarely happens that an animal ever dies from the disease, though when taken naturally it is not only very fatal, but, being contagious, spreads mischief among the neighbouring flocks.

Slipping the Lamb, and Protrusion of the Uterus.—Slipping the lamb is mostly caused by the ewes having been over-driven, worried by dogs when heavy with young, and sometimes by falling into ruts on their backs, whence they are unable to extricate themselves without assistance. It has also been attributed to feeding upon rape about Christmas time.

Protrusion of the uterus occasionally happens after an ewe has gone through a difficult labour, which has caused a portion of the uterus to be forced through the orifice. It should be returned to its place as soon as possible, but if not held in confinement by some means it is apt to protrude again. In order to keep it in its place some pass a piece of lead the size of a crow-quill through the lips of the orifice, and twist the ends to secure it. Others put a hog ring through the sides, or use twine for the same purpose. Either plan answers very well, and the ewe brings up her lamb without any further inconvenience. There is, however, a slight drawback to the use of the hog-iron, as it is liable to rust.

Sore Teats.—Sometimes the ewe has sore teats, and refuses to allow the lamb to suck. When this happens, the lamb should be put to another ewe if it can be possibly managed, or else fed twice a day with milk, either from a cow or taken by hand from an ewe. The udder should be bathed with tepid water, and afterwards washed with spirits or goulard, or a slight infusion of sugar of lead.

This treatment will generally be successful, but if not, and inflammation takes place, the part must be poulticed in order to cause suppuration.

Sore Udders.—After weaning, at times, the udders of ewes are affected by tumours, which need to be looked to, as they sometimes end in mortification. The part affected should be rubbed with camphorated spirits of wine, and if it suppurates it should be opened with a lancet, or other sharp instrument, and the wound cured by healing salve. It is better, however, to prevent this by milking the ewes, as previously recommended in weaning.

Sturdy, Dunt, Staggers, &c.—A disease known by these names, as well as by those of "Turn-sick," "Giddiness," "Goggles," "Sturdy-gig," and "Turn blob-whirl," is occasioned by a bladder in the head containing water, in which are "hydatids" or maggots, which find their way to the brain, the sheep mostly attacked by it being those under two years of age, after which time they are generally considered safe from its effects.

The animal appears stupid, as if deprived of its senses, and, instead of going forward in the usual method of progression, constantly turns round in one direction, in whatever side of the head the bladder may chance to be, and keeps away from the rest of the flock.

Cures have sometimes been effected by thrusting a wire up the nostril and destroying the bladder; or, if a soft place has been felt on the surface of the skull, the hydatid has been penetrated by means of a common awl, and in others trepanning has been successfully resorted to; but as more sheep die under these operations than are cured by them, it is generally thought best to confine the patient, and by cut food and oilcake get it ready for the butcher as fast as possible.

Ticks, or Sheep Lice.—Sheep, particularly if they are not in a healthy condition, are infested by a large tick (*Melophagus ovinus*), which seldom, however, occasion more than a temporary inconvenience. Mercurial ointment and tobacco water will eradicate them. Tobacco water should be made by boiling the tobacco in salt water. A wash of any weak mercurial preparation is effective. The invariable practice of dipping sheep and lambs each year, the former after they are shorn, is a good preventive.

Urinary Organs, Diseases of the.—Cystitis, or inflammation of the bladder, is rare with sheep, and is mostly confined to those which have been supplied with very nutritious food, such as beans and oilcake, and rams appear to be more liable to it than other sheep. Copious bleeding from the neck is recommended, followed by aperient medicine and opiates. A drachm of laudanum twice, or even three times, a day is a good remedy.

Worms.—Sheep rarely have worms in the intestines, but they have occasionally proved fatal to lambs, giving rise in the first place to a depraved appetite, which eventually causes irritation and inflammation of the intestines and fourth stomach.

It will be best to change the food, and if there is thought to be inflammation to give the following, mixed :

Linseed oil ... ... 2oz.
Opium (powdered) ... ... 3gr.

When the inflammation is reduced, in order to destroy the worms, the following should be given :

Linseed oil ... ... 2oz.
Oil of turpentine ... ... 4dr.

CHAPTER VI.

SHEEP (Continued).

Markets for Sheep—Driving to Market—Profit and Uses of Sheep—Importation of Sheep—Rearing Sheep Abroad—Technical Terms of Wool.

MARKETS FOR SHEEP.

THE markets for sheep are very abundant and numerous, not only where fixed markets and fairs are held, but, owing to the large consumption of mutton in all our large towns, there is no difficulty of disposing of either sheep or lambs, even butchers who are in a large way of business being now in the habit of buying a good number to suit the requirements of their trade. These they kill off as they want them, placing the sheep in what is called accommodation land, which often consists of a meadow in the immediate neighbourhood of a town or city.

Driving to Market.—As will be seen from what we have previously written on the diseases of sheep, they are apt to suffer from being over-driven. Unnecessary worrying by dogs, and blows, are invariably followed by painful results in one form or another, and the quality of the sheep, so to speak, deteriorates, and they will decrease in value in the eyes of would-be purchasers. The terror and the extra heat from which the poor animals suffer when driven with unnecessary violence, often encumbered with a heavy fleece, has an injurious effect upon the meat, and should always be avoided as much as possible, and feelings of humanity, as well as self-interest, point out that the animals should only be put in the charge of humane persons who will not abuse the trust confided to them.

PROFIT AND USES OF SHEEP.

The modern system of turnip husbandry and sheep-farming combined has had the effect of producing quite a revolution in the course of agriculture when compared with the old methods followed, and crops are raised with much greater rapidity than used to be the case from the benefit derived by the land after sheep have been placed upon it.

The direct results which are looked for, the value of the carcase and the wool, have been influenced very much in this country by their respective values.

At the close of the last century the finest wool manufactured in this country used to be obtained exclusively from Spain. Next to Spanish wool, the British short-woolled sheep furnished the best quality in Europe.

Great exertions were made to improve the quality of English wool, so that it should equal Spanish, and for that purpose Merino rams were imported and crossed with the best breeds of English sheep, especially as those known as the Ryelands, Mendip, Wiltshire, Southdown, Shropshire, &c., but although it was found the judicious crossings of these rams with ewes of the breeds mentioned resulted in wool even of the fourth descent being produced nearly as fine as the best Spanish, and great expectations were formed of the advantages to be derived, they were doomed to disappointment, for the Merino element in the carcase makes it unprofitable for the butcher.

A new competition also arose in the wool trade, in the shape of German wools, which began to be imported at low prices, Saxon and Bohemian wool superseding eventually the short wool of Britain in the manufacture of fine cloth in this country.

The improvements which were brought about in Saxon wool derived their origin from the same source ; the Elector Augustus Frederick, in 1765, procured 300 rams and ewes from Spain. and again, in 1778, imported 400 more. Application being made to the Crown of Spain, large numbers of the most celebrated breeds of sheep were sent over to England at the request of George III., who earned the name of the "farmer king," and these were widely distributed.

Although the experiment answered so well in Germany it failed here, the failure not being due to any inferiority of soil or climate, but to the high value of good mutton, which always realised a high price in England ; while on the Continent the case was quite different, the meat being comparatively of small account, while the wool would always command a ready sale, and experience has proved to demonstration that the improvement of fleece and carcase cannot be carried on *pari passu*, and that one must be sacrificed for the other.

Although, therefore, the value of English wool is below that of some others, in conjunction with the carcase it is still remunerative to the grower, the loss in price being counterbalanced by the increase of weight, the wool itself finding plenty of employment for inferior purposes in manufacturing.

IMPORTATION OF SHEEP.

Although considerable numbers of sheep are sent over to England from various continental ports, we are now comparatively independent of foreign supplies.

Admirable crosses of stock have now for a length of time been made in England, and these have caused the importation of foreign sheep to be looked upon as a matter of secondary importance. The original Herefordshire breed, called the Ryeland, to which we have just referred—a breed patient of hunger, and which will thrive on the scantiest fare—has been made larger in carcase and lengthened in wool by repeated crosses, mostly with Leicester, thus changing the character of the finest wool which England used to produce.

Even the original long-woolled sheep have been vastly improved. Leicestershire, one of the few counties which did not possess an early breed of short-woolled sheep, had a race of long, heavy, coarse-woolled animals, which, under the designation of the "Old Leicesters," have thus been humorously described by Marshall : "He has a frame remarkably loose and large, his bone heavy, his legs long and thick, terminating in great splay feet ; his chine, as well as his rump, sharp as a hatchet, his skin rattling on his ribs, and his handle resembling that of a skeleton wrapped in parchment."

Such was the unpromising raw material upon which breeders had to work, for, although crosses with the improved Leicesters have increased rather than have diminished the size of the original Leicester, yet the smallness of bone and the symmetry of form which the animals have acquired have considerably decreased the quantity of offal, and added very materially to the dead weight of marketable flesh.

Before the period of improvement set in, the mutton of these coarse sheep seldom amounted to more than half of their dead weight, whereas now the common average is more than two-thirds, and it is become an old saying that Dishley wethers, when well fattened, are in the proportion of an ounce of bone to a pound of flesh.

REARING SHEEP ABROAD.

Considerable attention has been paid to the rearing of sheep abroad, which has been done chiefly with an eye to the wool, and a good deal of pains has been taken with this matter in Germany especially, and in many European countries.

Many of the particulars which have been gathered together relative to the management, breeding, and rearing of sheep abroad, are very interesting, amongst which the rules promulgated in Fink's "Treatise on the Rearing of Sheep in Germany, and the Improvement of Coarse Wool" are noteworthy. These are to select, at the commencement of the undertaking, the finest woolled rams and ewes that can be obtained for the first generation, for if those for the second race be not finer than those used for the first, time will have been lost in effecting the proposed improvement. In like manner, to employ rams for the subsequent breeds quite equal to those for the first, or otherwise the intended improvement will be retarded. If an unimproved ewe be tupped by a ram of a mixed breed, and which has only one-fourth pure blood in him, the offspring will only have one-eighth of that race ; and by continuing to propagate in that manner, a complete separation of the two breeds will be at length effected.

While on this subject, we may as well quote the progressive scale of the amelioration of wool by the Spanish cross, given by Dr. Parry at the time when the merino sheep were exciting so much attention in this country.

The first cross of a new breed gives to the lamb half of the ram's blood, or 50 per cent., the second 75 ditto, the third 87 ditto, the fourth 93¾, at which period, it is said, that if the ewes have been judiciously selected, the difference of wool between the original stock and the mixed breed is scarcely to be discerned by the most able practitioners.

Although these hints are suggested for the improvement of wool only, yet the principle will be readily seen and applied in all breeding matters, whatever the object may be, for one can breed for any points that may be required.

The raising of sheep, however, in continental countries has been entirely dwarfed and thrown into the shade by the enormous proportions attained by the business in our antipodean colonies, and thousands upon thousands of bales of wool are now sent to us annually in the aggregate from New South Wales, Queensland, Port Phillip, New Zealand, Tasmania, Adelaide, and also the Cape—Cape wool being sold in large quantities at the Colonial Sale Rooms in London at every series of the wool sales, which attract large numbers of continental wool buyers—especially Germans and French—who keenly compete for all the best lots.

We must suppose that the meat of sheep finds its lowest level in the Argentine Republic, from a passage in Mrs. Brassey's entertaining book, "A Voyage in the Sunbeam," in which that lady describes a visit paid to an *estancia* two leagues across the Pampas, belonging to a Mr. Frer, who farms on an extensive scale, possessing 24,000 sheep and 500 horses, besides goodly herds of cattle. The authoress says : "The locusts have not visited this part of the country, and the pastures are in fine condition after the late rains, while the sheep look proportionately well. We passed a large *grasseria*, or place where sheep are killed at the rate of seven in a minute, and are skinned, cut up, and boiled down for tallow in an incredibly short space of time, the *residue of the meat being used in the furnace as fuel!*"

Technical Terms of Wool.

Each kind and class of wool at the colonial wool sales is carefully sorted and sold under distinctive descriptions, as fleece, scoured, skin, locks and pieces, greasy, &c., so that each buyer has the opportunity of obtaining the precise thing he may require.

The two main divisions of wool are those of the long-woolled breeds, which are termed "combing," and that of the short-woolled, called "carding." The combing sorts are chiefly used in the manufacture of baize, flannels, carpets, and other coarse kinds of goods, and takes its name from being passed in the course of its manipulation through combs that have upright steel teeth, the design of which is to separate the filaments and assort them evenly for the purpose of being spun; while the carding kinds, which are used in the manufacture of finer fabrics, are pulled crosswise between boards, termed "cards," which are furnished with crooked teeth composed of wire, which break it minutely into long rolls, which are spun in a certain manner.

The filaments of both, after their original properties have been designated, are called "pile," the fineness of which is gauged by the diameter of their fibre, while their length and strength constitute the "staple," and their serrated roughness denotes that "felting" quality which cloth manufacturers recognise as a material advantage in matting the cloth under the fulling mill.

This latter is peculiar to the short-woolled breeds. The freer the wool is from "kemps" or hairs, and the softer and more pliable it is, as well as possessing the necessary degree of elasticity, the more it is appreciated. It should also be uniformly white, and as free from stain as possible.

The chief points that are considered in the choice of wool are—freedom of the pile from "kemps" or "stickel hairs," which are of a brownish colour, and short, pointed, and brittle, and most often found in inferior breeds ; uniformity of strength throughout the fibre, and good even colour.   If the colour be bad, it is unsuited for manufacturing purposes where delicate shades are needed to be dyed, or for white goods ; the length of staple, as constituting the particular uses for which the wool is destined, that is to say, if long, for combing, if short and curly, for carding ; its degree of softness and elasticity, upon the union of which will depend its suitability for the manufacture of fine cloth ; its felting property, before alluded to, which, when combined with moisture, causes it to cohere, and forms a pliable compact fabric.

As wool varies considerably in its natural properties, as well as that each fleece contains portions of a different quality, the services of the wool-stapler have to be called in, who acts between the breeder and the manufacturer, it requiring great nicety of discrimination to separate the various kinds one from another.

When first shorn the fleece is usually sorted into three kinds. The "prime," which is separated from the neck and back ; the " seconds," which are taken from the tail and legs ; the " thirds," got from the breast and beneath the belly.

The same classification was familiar of old to the buyers of Spanish wool, the bags of which were separately marked " R.," " F.," " S.," which stood for *rafinos*, or prime ; *finos*, fine ; and *secundos*, seconds ; and occasionally " T.," for *terceras*, thirds or inferior ; but for a great many years there has been a much more discriminating classification, as many as fourteen sorts being separated by some, but commonly nine, as—prime ; choice and choice grey ; super and middle grey ; head, downright and third grey ; seconds ; abb ; livery ; britch ; and waste.

CHAPTER VI.

HORSES.

VARIETIES OF HORSES.

THERE are five distinctive breeds of farm horses in Great Britain and Ireland, and these may be said to be natives of the country. These are, the Lincoln, the Cleveland, the Suffolk, the Irish Garron, and the Clydesdale of Scotland. The latter have the credit of being reared in Lanarkshire, but they are found in all the southern parts of Scotland.

The Lincoln.—The Lincoln is the old black English horse crossed by Flemish, as they are generally termed, or Dutch and Friesland mares ; and are horses of great strength, fit for drawing heavy loads, and may be seen to perfection in some of the London brewers' drays. Notwithstanding their size and weight, these are often found trotting along through the streets of London at a smart pace, dragging a drayful of barrels or other goods.

It must be remarked, however, that the dray horse is not always black, as there are bays, a great many browns, as well as greys and roans.

H

In their natural state on a farm, they are slow; and smaller horses are generally thought to answer better in bearing fatigue, in being quicker steppers, and requiring less food, even proportionately to their size, and that, too, of an inferior quality.

There are farmers in Lincolnshire, and some of the neighbouring counties, who breed these horses with considerable profit. They are mostly sold as two-year old colts, the breeders retaining the mares for their own work, and for the purpose of breeding.

Those who purchase them, work them moderately till they are four years old, feeding them well during the time, and get them up into high condition as regards abundance of flesh to take the eye of the London brewers, for whose particular use most of these horses are bred.

Perhaps £40 may be paid for these colts at two years of age, and by the time they are four years they may fetch £80, making £40 profit, besides working moderately for two years.

But although this looks a paying matter at first sight, that it is really so has been very much questioned. It is assumed by the farmer that the horse earns the food he consumes during the two years; but this is considered very doubtful, for they rarely come to perfection till five or six years of age, during which time they must be treated very tenderly, and cost more than their work will repay.

It appears to be a matter of pride with the London brewers to use these ponderous animals, which are frequently seen 17 hands high, but, whether they are suited or not for this peculiar work, there is no doubt but that the smaller breeds are better adapted for ordinary farm operations.

It has been pointed out that there is a great drawback attending this large breed of horses, in the tendency, which they possess in a remarkable degree, to weak and convex feet; and to ossifications of the cartilages, and the pasterns. The predisposition to the former has been considered owing to their great weight pressing on the soft horn, induced by moist, fat pastures, and the latter most probably owing to the large amount of phosphates contained in the food which is given to them.

Farmers from a distance, who, not quite understanding this

branch of rural economy, have witnessed in the counties where these large horses are raised, four big animals slowly drawing a plough, have had their ridicule excited, not aware that these were simply young cattle being trained for work, and only being gently broken in.

The Suffolk-Punch.—The Suffolk-Punch is an extremely useful breed of horse that appears to have been kept tolerably pure for a great length of time, until a desire for improvement sprang up, and crosses were made that were not always judicious.

It has been described as being a large horse in a small compass, seldom exceeding 15½ hands high, and more frequently under. For farmers' use they are considered better than any other kind of horse in the southern and eastern counties of England. Steady going cobs for the use of elderly squires and country clergymen have been obtained from them, and although they can be met with in the original breed, a good deal of crossing at one time and another has been practised upon them. They are mostly chestnut colour, straight backed, broad and arched across the loins, with short couples, full and lengthy quarters, sinewy fore legs, and an open chest; the shoulder low, but well set for draught and the collar, but rather wanting depth in the chest, and being somewhat coarse headed. Inferior horses have ragged hips and goose rumps, but these are by no means common, and the breed generally is a great prize taker. Though horses of the pure breed are mostly of a chestnut colour, sorrel and bay are to be sometimes seen.

The old breed as a draught horse was (and is now when he is to be met with) unrivalled tugging along weights that seemed utterly disproportioned to his size, and, at the same time, doing the work cheerfully, with all the spirit and pluck of a thoroughbred, till his strength was completely exhausted.

It is often deplored by the surviving members of the last generation that the breed has been so crossed, that it is a matter of great difficulty to get a horse of the original pure breed.

The Cleveland.—This is a fine large breed of horses, measuring from 16 hands to 17 hands high, with a good dash of blood in them, and they are active, powerful animals.

In the old coaching days they used to be commonly used as heavy coach horses, but the numerous crosses made with the view of improving their speed have considerably altered the old strain.

From the prevalence of the colour they are commonly called Cleveland Bays, and though a good deal crossed, as we have remarked, they retain much of their original excellence. As they have a quick step, their value has been recognised in ploughing, when driven by whip-reins, though compact horses of smaller forms, are, perhaps, more desirable for the ordinary operations of husbandry. The breed originally was commonly found in every part of Yorkshire and the neighbouring counties.

The Clydesdale.—The Clydesdale is an ancient race, and is deservedly a favourite throughout the north, being extremely docile and steady, but the old breed is said to have been crossed by one of the Dukes of Hamilton, more than a couple of centuries ago, with Flanders stallions.

They are celebrated as draught horses, and may be seen in towns like Glasgow, dragging enormous loads behind them, their great development of shoulder and drawing power being very striking, 30cwt., besides the weight of the cart, is a common load for one horse, which traverses regularly twenty miles of road a day.

Mares are kept at moderate work to within a short time of their foaling, and the foals are weaned about the end of October. As a breed they require good feeding; they will work well, but they must be fed well, some of the Scotch coal-carters, it is said feeding their horses to the extent of a bushel of oats or beans daily. They do not get anything like this quantity of food with farmers, who give them from half a peck to a peck of corn daily, when doing ordinary farm jobs, but more if they are made to work hard, feeding them partly on potatoes, carrots, and a small portion of linseed, either in the form of oil or cake, given once a day in steamed food.

The Irish Garron.—The native Irish horse of the mountains is small in size, being about 14 hands, and light-limbed, but low in the shoulder, short-legged, with close pasterns, and he is very sure-footed. He is very hardy, but of no great value as a farm horse, and the attempts which have been made to improve the breed do not appear to have been successful, except towards the north, where some improvement has been effected by crossing with the Galloway breed, which is a stoutly-built horse, somewhat between a cart-horse and a saddle-horse. When long journeys were performed on horse-back, before stage-coaches and railroads, the Galloway was a breed much in fashion, and valued for his enduring qualities on the road.

The native Irish horse is hardy and indefatigable, though often he has to put up with very bare fare, and he makes a capital roadster. A good many horses of one sort and another come from Ireland, but it has been used against many of the farmers there as a matter of reproach, that almost everyone who has 100 acres under tillage keeps one or two breeding mares, which are worked to within a fortnight or so of the time for dropping their foals, the colts being mostly sold at three years old, and that the only qualification that is thought of regarding the stock is the size of the sire and the price of the covering, which is seldom to exceed three half-crowns, or, at most, half a barrel of oats. The system upon which matters used to be conducted, it is said, has prevented improvement, shambling blood horses being used to cross the native stock, which has produced a race of mongrels.

It cannot, however, be a matter of surprise that small Irish farmers desire to add to their gains by bringing up a few horses annually, which many manage to do at a small cost, and by following economical methods of feeding.

REARING FOALS.

As to the desirability of rearing foals on a farm, Mr. Hooper, in a paper read before the Ballineen Farmers' Club, says : "The farmer may not have all the appliances and arrangements of a stud farm, with its paddock, open yards, loose boxes, &c. ; still, he may like to rear a foal or two every year, and, without entering deeply into the question whether it pays to breed farm horses, I think most will agree with me that it is better for a farmer to have horses to sell than horses to buy. Moreover, few farmers can afford to give up the whole time of their brood mares to breeding, nor do I see any reason why they should. Neither the mare nor the foal she is carrying will be any the worse for her regular work on the farm, provided she is well fed and not put to extra hard work, nor to work to which she has been unaccustomed. Of course, the nearer the mare gets to her time of foaling the lighter must be the work required of her, and a week or two before the time she should be left perfectly idle, especially if not in high condition. If the mare foals very early in the season, and the weather is bad, she and the foal should be put in a loose box at night, and allowed as much liberty by day as the weather will permit ; but if she does not foal before the month of April, she may be kept in a sheltered field day and night, and, unless the pasture is very good, a feed of oats should be given her every morning. Now, we all know that the foal would be better if his dam had nothing to do but to suckle him until he is old enough to wean ; but I am speaking of rearing horses on farms where the work of the farm is of primary, and the rearing of foals of secondary, importance. In general, then, the mare may be put to work a fortnight after foaling, and when she is at work, the foal should be shut up in a light, airy, loose box, and the mare taken in to suckle him at intervals of not more than two hours. Of course, the mare must be highly fed when doing this double work, and should have as much green food as possible. In addition to the dam's milk, I always give the foals of my working mares a quart of cow's milk, brought to a natural heat—a pint about an hour and a half after the mare has gone to work in the morning, and a pint in the afternoon. At first the young thing is very shy, and is frightened at being caught and having his nose held into the milk ; but he soon learns to expect it, and will come and drink it out of the vessel when held to him. When the turnips are all sown and there is very little work for the horses, the foal can get two or three months of uninterrupted liberty with his dam. I find the most convenient time to wean is when the working horses are brought into the stable at night. By that means the foal is accustomed to the company of the other colts (or a donkey will answer the purpose, if there are no other colts on the farm), and will stay there quietly without his dam. In about another month (according to the season) I bring him in also at night, putting him with another of the same age, or not more than a year older, into a loose box, and giving them as much chopped furze and hay as they will eat, and three or four pounds of crushed oats; and here let me remark, there is no time of his life at which a horse gives a better return for a few oats than the first winter. I turn them out the whole day in all weathers, excepting very hard frost or snow, and put them out altogether as soon as the weather and grass are good enough. The feeding the second winter is merely a repetition of the first."

Training Horses for Farm Work.

It must be admitted that, as a rule, the great intelligence possessed by the horse is not appreciated and made the most of by farmers in general.

Next to early working and bad and careless feeding, the bad breaking-in of horses is much to be deprecated, for, unfortunately, amongst farmers, the common way is to put the horse in the harrow when he is three years old, and, should he be lively and spirited, work him down; if stubborn or sulky, flog him unmercifully, very often about the head, recourse being seldom had to gentle means of training.

By this course of usage the animal grows vicious, his temper being ruined by ill-treatment, when with proper usage his natural intelligence would be considerably more developed, and he would strain every nerve in the service of his master. Of the attachment and fidelity of the horse there have been given many remarkable instances; and young horses, instead of being used with severity, should be coaxed to their work.

If kind usage and gentle means fail with an intractable horse, harsher measures may then be adopted, but these should be used with temper. With horses it will be found that, in nine cases out of ten, gentleness succeeds better than severity.

It will be found the best mode of training horses to farm
work, after accustoming them to be handled and the sound of
the attendant's voice, to place them in the plough with a steady
old horse, under the care of a sober, good-tempered, intelligent
ploughman.

When two years old off they may be brought into use, or in the
course of that summer, but they must not be worked too much,
and only for half-days at a time, for their immaturity will not
allow them to bear greater exertion until they have reached the
age of four years. By the gradual increase of their labour they
will then be brought to the full extent of their powers, and their
strength may be fully employed without injury, provided they are
not made to exert it too much by taxing their speed. In the
ordinary way a working horse will seldom suffer by any fairly
managed labour, provided it is leisurely executed—that is to say,
without any very material increase of their usual and natural
walking pace.

ADVICE IN BUYING.

As the subjects treated upon in the present work are those
which relate to farm operations, it does not come within its
scope to speak of the various horses reared for special or definite
purposes, such as racehorses, carriage-horses, &c., except in a
cursory manner; but it will be quite in place if we make
slight reference to saddle horses and hackneys, which an agriculturist may either require for his own use or may have an
opportunity of purchasing, which he may do either with the
view of keeping it permanently or of improving its condition
and selling it again to an advantage.

Good carriage horses and good hunters are to be had easily
enough by those who are willing to pay the price for them, but
to get a good hack is a much more difficult matter, because an
union of excellent qualities is necessary to make up a good
hackney.

He should be good-looking, strong, and well bred; be safe and
sound, an easy and fast trotter, and a good walker—that is to
say, he should be able to walk four miles an hour and trot ten

within the same limit. Should he be required either to canter or gallop, though these are not the lines he is expected to shine in, he ought to be able to do either and command a certain amount of speed.

While, on the one hand, he must not be lazy, on the other, he must not be skittish or restive, and neither shy, stop short, nor turn round. It will be seen that to possess all these diversified qualities, which make up the sum and substance of a good hackney, requires an animal which will be somewhat difficult to obtain.

Where it can be done, it will be found the best plan to look at a horse first, as he stands in the stable, which will give an opportunity of noticing whether he is gentle to approach and free from vice; an opportunity will be given also of inspecting his forelegs, and whether he points or favours either of his feet, or shakes or knuckles at his knees or fetlocks, or stands with his legs too much under his body. Either of these peculiarities must be counted as defects and drawbacks to the horse.

The eyes and teeth next require examination. If the eyes are small, shrunken, cloudy, or contain large specks, these are grave faults, which indicate that the sight is defective, though sometimes the outward appearance is not always to be trusted alone; for horses which have become blind through the optic nerve being paralysed have at times exhibited clear transparent eyes. If the teeth are very long, or project horizontally from the lower jaw, or have three corned faces, it is a proof that the animal is an old one. On the other hand, if some of the teeth are smaller and whiter than the others, he is perhaps too young.

The horse on being led from the stable should not be allowed to be hurried or driven, but the intending purchaser should take care that he is walked and trotted gently, without being excited either by fear or emulation, by which means his natural action may be seen, and whether he stumbles or goes lame, violent action when driven concealing the last-named defects very often.

The horse should next be mounted, previous observation having been taken of his general appearance, height and strength, and the state of his limbs and feet. At first he should be walked carefully in order to ascertain whether he walks fast and safe, and if he shies or is restive. The pace should then be increased in order to discover whether the animal is sluggish or impetuous, ending by a gallop, to find out if his wind be sound or if he is a roarer. A good hack should be free from excess of muscular development, with good, deep, well developed shoulders.

In buying a cart mare with an eye to breeding, choice should be made of one long and roomy in proportion to her height, and full in the flank, signs which are a promise of her becoming a good nurse. Her temper should be gentle, and she should be possessed of a sound constitution ; she should be free from hereditary defects, and have those desirable points that are wanted to be transmitted to her progeny.

Accommodation for Horses.

In some old homesteads the teams are often huddled together into low-roofed, narrow, dark, and unstalled buildings, where fumes arise from stagnant urine and accumulating heaps of fermenting litter, the consequence being that though the horses may be sleek from good feeding, their coats are foul and their heels greasy.

A stable for farm horses should be roomy, clean, and well ventilated, and each horse should have from 1200 to 1300 cubical feet of space allotted to him. The French put the requirements of an ox or a horse below this, calculating that if there are 800 to 900 cubic feet of space in a properly-ventilated building for each it will be sufficient.

These matters are now much better understood than they used to be, and the first conditions which are insisted upon with stables is that the whole site of the buildings be thoroughly drained and that good ventilation be secured. The absence of an upper story or loft is desirable, but if there is one, there should be openings in the roof to allow of the escape of the heated air. Plenty of light is desirable and necessary, so that dirt may not be overlooked, and due provision must be made for the removal of urine and excrement in the most complete manner, and as speedily as possible.

Each horse should have his own stall, and, for horses of the ordinary size, 6ft. is considered the proper width, but for small horses 5ft. 6in. is sufficient—the object being to have it of such dimensions that the horse cannot fairly turn himself round in it, while yet not being sufficiently narrow as to cause the animal to be uncomfortable in it.

It is desirable to have a non-absorbent surface to form the floor of the stable, which should be so pitched as to allow the urine to flow from off it easily. With this object in view, many floorings of stables are made to slope 6in. from the front wall.

This steep slope is, however, by no means necessary, and in the first place it causes the bedding to be kept up with difficulty, and next it is sometimes the occasion of injury to the horses, by straining their hind quarters.

To avoid these results, as the staling of a horse only reaches about half way up the stall, if the floor from that point is made to slope very gently from the sides towards the centre, the water is much more easily carried away than by any other manner, a fall of a couple of inches being all that is necessary. A channel for the water to pass off is generally also formed at the horse's heels, extending along the stable at about 10ft. distance from the front wall.

Underneath the channel a drain pipe should be laid, having an eye provided with a trap, and a grating at the centre of each stall.

Some good practical farmers, particularly in the county of Suffolk, keep their teams, not in stables, but entirely in open yards, or " hammels," surrounded with well-littered sheds for them to go under at will, and it has been proved by experience that their general health may be as well maintained in this manner, if not better, than in stables. In the eastern district of Suffolk it used to be the practice not to allow the horses to remain in the stable at night, but to turn them out when fed in the evening, by which means they became hardy, and were neither subject to swelled legs nor to cold and inflammation,

which are often caught through changes of the atmosphere,
and a horse being removed from a warm place to stand for a
time in a cold air.

These yards are made to do duty during the whole year, for
summer soiling and winter feeding; there are, however, three very
principal drawbacks to this plan. The first, that horses are
somewhat exposed to accidents when a number are thus put up
loose together; the second, that their food cannot be so equally
proportioned to each separate animal; and the third, that it is
not possible to keep them equally clean.

Labour Required for Superintending Horses.

In the stables of gentlemen, where the horses are kept
remarkably neat, a good deal of labour is often expended, but
common working horses require only a little grooming, yet their
coats must be kept clear of scurf, and their feet should be well
attended to. In some counties the rough hair which encumbers
their fetlocks is a useful protection against flints, but a much
less quantity would often serve this purpose than what is allowed
to remain on, which, when clogged with dirt, is apt to bring on
grease.

There are some carters who entertain such an affection for the
animals they are associated with that they would rather starve
themselves than their horses, though there are others, on the
other hand, who are only too prone to neglect them. Such
men are, however, unfitted for their occupation. Excess of care
and regard, however, producing mistaken kindness, is apt to cause
an injury to the animals, for an affectionate carter often does
not hesitate to pilfer corn, thinking the best thing he can do to
his charges is to cram them to the utmost.

This extra care, however, both on the score of health and
economy, ought to be discouraged. The chaff as well as the corn
should be weighed and measured, and even the hay, where it is
commonly kept loose, should be trussed, so as to enable it to be
served out accurately. The expense and trouble of binding will
be more than repaid by the economy of consumption and saving

of waste. It has been justly observed that in stinting the quantity of food given to horses—so that it is not unduly done—causes the men to become more careful ; they look upon it as something, and know that if they use it extravagantly to-day they will run short on the morrow, when the cattle are allowanced. The servant by this means learns to be frugal, while the cattle are fed regularly, and he will give them a little at a time, so that they eat it up completely, and it is not blown over and wasted.

FEEDING.

Corn is the most nourishing food that can be given to horses, and oats are generally considered as the best adapted to their constitution. Close observers have, however, remarked that grain of other species, when mixed with a proportionate quantity of straw chopped fine, or bran, to supply the place of the husk of the oats—without which no corn should be given—effects the same results when supplied in the same relative proportion as to weight.

It may happen that a farmer may have grain by him other than oats which it would be desirable for him to use in feeding his horses, rather than be money out of pocket by purchasing oats. Horses, when not hungry, will endeavour to separate corn so mixed from the chaff or bran, and to prevent this it is customary to sprinkle the food with water. Should this course be followed, the greatest care should be taken to have the manger thoroughly cleansed, for nothing is so offensive as such food when it has become stale.

It is essential that grain of every description should be in a perfectly dry state to cause it to be fit to feed horses with, and free from the condition of fermentation which arises from dampness. In a wet season, oats which have been harvested in bad condition have been known to occasion epidemical disorders amongst cattle, and by giving oats to their horses too soon after they have been reaped, farmers have inflicted severe injury upon them, though corn which has sprouted is not unwholesome, provided it has not heated and has not a bad smell ; and barley

which has been slightly steeped for two or three days, without
being dry malted, is considered particularly nutritive.

The stomach of the horse is constituted in an entirely different
manner to that of the ox; and while the latter can dispose of a
considerable quantity of food at a meal, the former is adapted to
consume only a moderate quantity, and frequently the horse's
stomach contains about three gallons, whilst the ox has four
stomachs, the first of which is considerably larger than that of
the horse.

The confined space of the horse's stomach would seem to indicate that he was intended by nature to consume grain, the
formation of the molar teeth confirming this opinion. Poor,
bulky food, such as ordinary roots, is not, strictly speaking, suitable to the horse, though it is very often resorted to for reasons
of economy, even carrots, which are the best of them, and which
are eaten readily, containing eighty-five per cent. of water.

These are a capital aid, but they should not be depended upon
alone, but used conjointly with more concentrated food. Good
hay is the cheapest dry food which can be given to horses, when
the nutriment it contains is taken into account, but is too bulky
by itself. Oats are reckoned to be one-third dearer than hay,
when given in an equivalent quantity.

Beans are more concentrated than oats, and form a capital
staple of food as regards their flesh-making properties, and
they are also cheaper ; but they are open to this objection, that
if given too freely, being very heating and stimulating, they are
apt to produce inflammatory swellings of the limbs. They are,
however, given very advantageously in conjunction with oats
when horses are heavily worked, one-third to two-thirds of oats
being about the best proportion to give. Some use a larger
proportion—as much as a half—but as beans are astringent in
their nature, when they are cheap and oats dear, and the
temptation for economy's sake leads to the somewhat free use
of beans, it will be advisable to give bran with them also,
bran being relaxing in its effects. The two, however, are not,
even when combined, entirely capable of supplying the absence
of oats, as both are deficient in starch.

When roots and cheap substitutes for corn are given, the food should be steamed or cooked. It assists the masticating powers of old horses and is a safeguard against ravenous feeding. To each feed of steamed food 4oz. of common salt should be added.

Peas are sometimes given to farm horses without any prejudicial effect, but they should not be given in summer, at which season they are found to be too heating. They should also be at least twelve months old, and at that age what they have shrunk up in quantity will be made up for in quality.

In feeding horses, it is best, as a rule, to give concentrated food; for although it has been observed that when a large quantity of hay has been given, with the view of minimising the quantity of corn, although horses will acquire more fat than upon the same proportion of grain, yet they are not capable of doing so much hard work. On the other hand, when the hay is diminished and the corn increased, although, perhaps, the horses may get absolutely thin, yet the flesh they have upon them is firm, and they are capable of more continued exertion.

It has been estimated that a horse consumes one-eighth less of meadow hay than of that made from the artificial grasses, and that the different forms or qualities of hay stand in the following relation to one another: In the ordinary way, 8lb. of meadow hay are equal in support and nourishment to 3lb. of oats; 7lb. of saintfoin, tares, clover, or other succulent hay, are supposed to be equivalent to the same quantity, saintfoin standing the highest in the scale, but 9lb. of hay will be required if made from poor pastures.

Old hay is much better than new for horses; the longer it stands in the stack the better it is, up to a certain point, that of one year of age being very wholesome. The second growth, or after-math, is not so nourishing, and should be given to the cows where there are any. In order to preserve the freshness of the aroma of the hay for horses, it is desirable to get it into the stack as soon as it is properly dried.

Summer Feeding.—In old times the summer feeding of horses was almost entirely confined to pasturage, but of late years the

advantages of soiling upon green food horses which have not very hard work to do has become very general.

There are both advantages and disadvantages connected with grazing. A chief benefit consists in its requiring but little attendance, and, being the most natural form, is consequently the most healthful, and is therefore to be preferred for all young cattle which can be spared from constant labour. But for horses at regular work there are several disadvantages connected with it. These are the loss of time in getting them up from the field; the indisposition to work, of which they acquire a habit, from being at large; their being subject to suffer from the heat, or being tormented by flies; the waste and injury done to the grass if it be valuable land, by their trampling upon it with their iron-shod feet, and the loss in the value of their dung and urine.

On the other hand, from the practice of soiling, there is a more economical consumption of grass, whether it be natural, which is mown for them, or artificial, than by grazing. There is a satisfactory accumulation of manure ready to be placed on the land wherever it may be needed, and the cattle enjoy cool and quiet in the midday heat beneath sheds or in stables, where they are always ready when wanted without a man's having an exciting chase after them with a halter in one hand and a sieve of oats in the other.

As, during summer, there is a long interval of rest between the morning and afternoon work generally, the carters can in this time cut the necessary quantity of food that is required for use.

It is a good plan to partially follow both systems in the summer feeding of horses, so that they may be both grazed out and soiled. They may be fed upon cut clover, or other artificial grasses, in the middle of the day, between yokings, and turned out at night into well sheltered meadows or inclosures. They thus enjoy the advantage of being under cover during the heat of the day, when they will be sheltered from flies, and will be able to feed at their ease, while from being pastured out during the night they will be able to select those herbs and grasses which instinct teaches them is necessary for their health. All animals prefer a variety of grasses, which they cannot obtain when soiled. Exposure to the night air in summer time is also highly favourable to their health.

Lucerne is capital feed for soiling purposes, but as it requires a peculiar quality of land to grow it, it cannot universally be cultivated. Tares also come in very useful—in the first place, because the winter species is ready earlier than any others, excepting rye grass, and furnishes a weightier crop. Commonly, tares are ready for cutting by the middle of May, for, if left till they are quite ripe, they will become unfit for soiling before the crop is completely used ; and though there may be some loss in using them at the commencement, it will be made up by the saving at the latter end ; they have the great advantage of giving the opportunity of beginning the soiling earlier than it could be otherwise done.

Before they are off clover will be ready, and the tares then standing can be made into hay. In its turn, again, by the time the clover begins to get strawey and loses its succulence, spring tares will come in, which, if they have been sown at intervals of about a fortnight, will last until the second cut of clover is ready.

In some parts of the country a very good system is followed of saving a portion of rye alone, to be cut green, then another portion of rye with tares, and afterwards the remainder entirely with tares. The rye comes first into use, and assists in raising the earlier tares, while those of later growth do not require such nursing.

A succession of green herbage may thus be secured for the horses for four months throughout the summer and autumn, without having recourse to the meadows at all, the horses requiring, while thus fed, but very little corn. Indeed, horses can be fed and kept in fair condition and worked throughout the summer without any corn at all, yet the possibility of this being done in no wise interferes with our previous views, as to the apparent intention of Nature with regard to what constitutes the fittest kind of food for a horse.

It is a good plan to give some green food along with the corn and chaff, before the usual period comes round of feeding entirely

I

on dry fodder, so that the change from green to dry and dry to green should be gradual.

It is a well understood principle amongst thoroughly experienced persons who have made this subject a study, that the custom of giving corn along with green food is unprofitable. The grain when thus mixed passes rapidly off the stomach, and is never thoroughly digested.

Corn is prescribed to be given to those horses whose increased labour and exertion demand an addition of substantial food, but that corn should be given only in the morning and at night, accompanied with sufficient chaff to give it the necessary consistence, and the green food should only be given at mid-day. This precaution is very frequently omitted by numbers of people, and it is very common with them to allow half the usual quantity of corn, without taking into account the effect of the watery juices of the green food upon the digestion, by which means a great portion of the nutriment contained in the grain is thrown away, as it were. On the other hand, when horses have been fed exclusively upon dry food, a little green is imperatively called for.

Summer Feeding in Holland and Flanders.—According to Radcliff's report of the agriculture of East and West Flanders, the practice of summer soiling there must extend over a considerable time, the feed of the horses being limited to half an acre of meadow grass, cut and carried to the stables, from the middle of May till the middle of June, from which time till the end of August one-sixth of an acre of clover is added with 2lb. daily of beans; and thence to November, when the winter feeding commences, the clover is replaced by an equal quantity of carrots.

A farm of 200 acres is mentioned in the report as being cultivated by eight horses, each of which gets daily in winter 15lb. of hay, 10lb. of straw, and 8lb. of oats, and, after every feed, a bucket of water richly whitened with rye or oatmeal. In summer clover is substituted for hay, but the other feeding remains the same, and the "white water" is never omitted.

Feeding Horses in Scotland.—In Scotland, well-washed turnips and potatoes are given steamed and mixed with sifted wheat chaff,

oat husks being avoided, as they are thought to be injurious.
62lb. weight of unsteamed turnips and potatoes produce 47lb.
steamed. To each feed of steamed food 4oz. of common salt are
added, mixed up with one-fourth part of a bushel of wheat chaff,
weighing about 1½lb., a greater quantity of wheat chaff being
thought to have a laxative effect. In spring, bruised beans and
oats are given instead of bruised oats alone. From June to
October, when horses are doing heavy work, such as drawing
manure, harvest work, or working the green crops, they are fed
with cut grass and tares in the house, and about 7lb. of oats
each day given at two meals, and increasing or decreasing the
quantity according to the work they have to do. Good Scotch
farmers disagree with turning out those horses to grass which
are regularly worked, considering that the changes in the climate
often lay the foundation of diseases which they escape by being
housed.

Method of Feeding Horses in Ireland.—Upon the occasion of
Mr. Hooper reading a paper on the rearing and feeding of live
stock, at a meeting of the Ballineen Farmers' Club, he gave
an interesting account of the method he pursued in feeding
his horses, which is well worthy of notice, for the practical
details which are given. "A farmer," says Mr. Hooper, "not
only requires to feed his horses well, so as to keep them up to
their work, but also economically. To this end it is indispen-
sable, in my opinion, that he possesses two machines, viz., the
furze cutter and oat bruiser ; the price of the hay and oats saved
will soon repay the cost of both. I find that my horses do as
well upon 10½lb. of crushed oats as on 14lb. of whole oats.
They are fed in the following manner : From the time they are
put to regular work after harvest and stabled at night they get
3½lb. of crushed oats the first thing in the morning (at five o'clock
in the autumn and spring months, when they go to work at half-
past six ; and at six, or a little before it, in the dark winter
mornings, when they cannot work till after seven), about an hour
and a half before they go to work. At twelve o'clock they get
another 3½lb. of oats and a pannier of chopped furze, and at night
another 3½lb. of oats, and two full panniers of furze. On this

feeding I have kept several horses in really good working condition all this winter; but I have some horses (not of my own rearing) which will not eat enough furze, and to them we give a little hay in addition. About the end of March the furze becomes less nutritious, and we substitute hay for it ; and, as the days are then long and the work hard, we give, in addition to the three feeds of oats, a small bucket full of boiled swedes, mixed with a little meal or bran, to each horse at eight o'clock in the evening. I do not allow the cost of the meal or bran to exceed 1s. per week for each horse, so that the food, including turnips and firing, costs something less than a feed of oats. If the turnips are finished before we can cut any grass or vetches, I give a feed of boiled barley instead ; this costs as much as a feed of oats exclusive of the firing. As soon as grass or vetches can be substituted for the hay, I find two feeds of oats sufficient without any boiled food. As soon as the turnips are all sown, I turn my horses out, and as there is then very little work to be done, and only one, or at most two, are required to be at work at the same time till harvest, I give them no more oats. This run out to grass during the summer months I consider is essential to the horse's health.

"The cost of each horse per annum will thus be as follows: 10½lb. of oats a day, reckoning the oats at the extreme price of 1s. a stone, will be 9d., i.e., 5s. 3d. a week, or £6 16s. 6d. for twenty-six weeks. The cost of the furze is merely the rent of the land on which it grows, as there is no expense attending its cultivation.; the cost of preparing it is about 1s. a horse per week ; this gives 26s. to be added to the £6 16s. 6d., making the cost of feeding for twenty six weeks, from October to March, £8 2s. 6d. We now come to eight weeks during which the horse is fed on hay, and has an extra feed of boiled food at night, to be reckoned at the same price as a feed of oats. This will give 1s. a day, then, for four feeds, i.e., 7s. per week, £2 16s. for the eight weeks. As for the hay, I must confess I have never weighed it, but I think we may put it at 1cwt. and a quarter a week— say, 4s. ; this gives 32s. to be added to the £2 16s., making £4 8s. for eight weeks, from the beginning of April to the end May. We have still some four weeks left, during which the horse requires two feeds of oats a day ; these, at 6d. a day, will amount to 14s., and reckoning the cut grass at 3s. per week, will make 12s., or £1 6s. for the four weeks' feeding, During the remaining fourteen weeks of the year the horse gets no corn, except a few sheaves of oats in harvest time, and perhaps a feed of corn now and then, if he has to go a journey, which I think need not be reckoned in the cost of his keep. Valuing his pasturage, then, at 2s. 6d. per week, we have the sum of £1 15s. for the remaining fourteen weeks' keep. We have, then, £8 2s. 6d. for wenty-six weeks, £4 8s. for eight weeks, £1 6s. for four weeks, and £1 15s. for fourteen weeks' pasturage, giving the cost of each horse for the year £15 11s. 6d. To this must be added something for the furze (a horse will eat about a quarter of an acre) and something for the wear and tear of the furze cutter and oat bruiser (I have had mine in constant use for ten years, and they are doing their work as well as ever). With whole oats the cost would be £4 more on each horse per annum."

Why Horses Eat Dirt.—It has been pointed out by a writer that when horses have been turned out of badly ventilated, close stables, they may often be observed to lick up the dust in the roadway, which they swallow with apparent relish. When this is seen, it is a sure indication of acidity of the stomach, and Nature has given the animal the instinct to endeavour to cure this by appropriating those alkaline qualities found in the earth, and of which the stomach is deficient. Like the pig, which may be often seen eating cinder-ashes with avidity for the same reason, horses are at a disadvantage in the matter of stomach complaints, inasmuch as they can neither rid themselves of anything disagreeable by vomiting, like the dog, nor can they belch up accumulated gases. The stomach of the horse is too small to admit of heavy feeding, which tends to the formation of gases, which cause distension, and hence the necessity of both careful feeding and of allowing them to take in by inhalation in a well-ventilated stable the pure air, which tends to counteract the impure. Like human beings who suffer from acidity, or heartburn, to whom a dose of soda or magnesia is useful, they lick the lime from the walls or the dirt from the roadway. For this

reason a lump of rock-salt and another of chalk may be placed
with advantage in a corner of the manger, and an ounce of
common salt put in the water daily will be found to be of
service. Sometimes this disease or vitiated appetite may be
caused not by anything defective in stable accommodation or bad
feeding, but, according to recent opinions expressed by chemical
authorities, may arise from a deficiency of mineral properties
in the food, which has been grown upon poor soils, deficient
in silica.

This may be corrected by administering a course of tonics.
For this purpose take of oxide of iron, 2oz.; powdered nux
vomica, 1½ drachms; powdered fennel seed, 2oz. Mix these
ingredients together, and divide into eight powders, one to be
given at morning and one at night. The best plan, however,
to obviate the use of medicine is to pay great attention to proper
feeding, and for this purpose sound hay in moderate quantities
at regular intervals will be found very advantageous. It is also
desirable to give a change of food in the form of carrots,
turnips, or other roots. When horses have passed through a
winter chiefly fed upon dry food, there is often occasion for
administering a little medicine, which there would be no
occasion to give had they been fed upon some green food, which
is obtainable as the spring approaches. A good spring medicine
for a horse is 1oz. nitre, 1oz. of sulphur in a bran mash,
mixed up well, and while warm to be covered over with a cloth.

We merely make cursory mention of this recipe here, apart
from regular medical treatment, which we shall speak of at
greater length.

Winter Feeding.—In much the same way as summer feeding,
working horses can be supported during the autumn and winter
with only roots and plenty of hay and straw (without giving
them any corn) until the young grasses make their appearance.
Yet, although they can be maintained in health and apparent
vigour, they will not be capable of any strong or unusual exertion.

Raw carrots washed and sliced, without the tops, are most
relished by them, and they invariably eat them in preference

to any other roots. Cases are not uncommon where animals have been known to refuse even to taste any others, either cooked or uncooked. This has probably arisen from want of care in the preparation, they being perhaps boiled too soft, or being given in a mashed or wet state.

This, however, can hardly be a matter of surprise, for some little management is necessary in order to get horses to take to a kind of food with which they are unacquainted, but they may be got into the way of doing this by offering them a small quantity of any new species of root by hand, when there is nothing else to eat, and to apparently coax them. When they can be got to relish a little in this way, a small quantity should be mixed with their ordinary food, and by this means their dislike can be gradually overcome.

Most animals derive support from their solid food, and in accordance with the care with which it is chewed and masticated, and this fact has been recognised for a long time, for even before it was generally usual to convert provender into manger-meat in this country, this practice was universal in Flanders, where all the farm horses are in high condition.

There is a great advantage in giving ready cut and prepared food, by which digestion is assisted, for old horses lose much of the power of mastication, while young and greedy cattle very often swallow a considerable part of their corn entire, when it is given alone; which, passing through them in the same form in which it was given, affords no kind of nourishment.

Giving an unlimited quantity of rack meat to horses is often a mistaken kindness, for not only will some of them pass whole nights in eating, when rest would be doing them much more good, but by the unnatural distension of the stomach, its powers are weakened and the general health of the animal is injured.

Many farmers now have their stables fitted up without racks at all, but the food cut up and mixed is placed in the manger, and given several times during the day. By this means a more thorough mastication is insured and digestion is assisted. The food is eaten in less time, and there is an advantage, too, in it, that damaged hay or straw, which would be refused separately, gets consumed with the rest. It can be more readily weighed and measured, and be more accurately distributed to each horse. There is not any waste, and consequently the food goes further.

Some of the best horse feeders in winter pursue this plan. Every particle of food is given in the form of manger meat. The hay and straw are cut very short, the beans are bruised, and the oats sometimes coarsely ground. When barley is given, it has been found most advantageous to wet it, and allow it two or three days to sprout. The chaff is cut by a horse-mill, which, at the same time, bruises the beans by a small addition of machinery, and one horse with a couple of lads, one to unbind and deliver the hay and the other to fill the troughs, will cut a load of hay in three hours.

Nothing more is needed in the way of machinery than a common chaff-cutter of somewhat larger size than those used by hand, and a bruising apparatus of very simple construction, both of which may be readily added to a thrashing-mill. It can work in a loft above the horse-course, on the floor of which the whole provender is afterwards mixed, the chaff being spread first, next the bruised corn, and afterwards bran, when it is used, and the whole united in one mass. When used it is measured off in distinct portions.

The usual allowance of hay and oats to riding horses standing in stables and doing occasional work only is 12lb. of hay and 10lb. of oats, but a common hard-working cart-horse cannot be supported in good working order upon less than 28lb. or 30lb., or more if straw is substituted for any portion of the hay.

In some of the brewers' and coal merchants' stables, where horses of large size are employed, which have to drag heavy loads and exert a great deal of muscular power, 36lb. to 40lb. of dry food is sometimes given daily in the proportion of 16lb. of clover or saintfoin-hay, and 4lb. of straw, to 18lb. of grain, either oats, barley, beans, or peas, and 2lb. of bran, oats being preferred to beans in summer, but more bran is given with the latter than with the former.

Grains are not a good thing to give horses, however desirable for cows, which relish them exceedingly. It is said that the horses of brewers, to which in some instances they have been copiously given, so fed become rotten, and die in a few years, and on dissection are found to have large stony concretions in their bladders.

Feeding for Exhibition.—From the numerous hints which have been already given enough can be gleaned to show how horses can be kept in the best condition, but for the purpose of exhibition, it is needless to say the first quality of food should be supplied, and an exception made in those economical expedients which on the face of them show they are not calculated to improve either the condition or appearance of the horse, however desirable their practice might be on the score of economy.

A horse, unlike an ox or a pig, does not require to be made too fat, but the first essential is to keep him in perfect health with sufficient exercise, abundant grooming, and enough of the best food, occasionally a little varied.

GROOMING AND EXERCISE.

Grooming.—Farm horses do not require very much grooming, but it is highly necessary that their skins be kept thoroughly clean. On the other hand, the inmates of the gentleman's stable are expected to be seen with their coats shining both from the effects of perfect health and the liberal use of the currycomb and straw wisp. Not only on account of the animal's appearance, but on the score of health, great attention should be paid to this point. Horses are subject to the attacks of insects, not only internally but externally, and the parasites which inhabit the skin are only to be eradicated by care and good grooming. A louse, called *Trichodectes equi*, sometimes infests them, especially after pasturing. These may be removed by an ointment composed of white hellebore, flowers of sulphur, oil of tar, and train oil. The itch also consists of an almost invisible mite, termed *Acarus scabies*, which lives beneath the skin. Good food, good grooming, and general attention will keep the skin of the horse free from these visitations.

Exercise.—The daily work of a cart horse or animal employed in agricultural labour is generally the means of affording him as much exercise as he wants, and, in the case of somewhat lazy animals, perhaps more ; but riding horses that are only irregularly employed, and at times have to bear considerable fatigue at a rapid pace on occasional journeys, should always have a fair amount of daily exercise to keep them in equal and invariable health. Without exercise the muscle and viscera are not sustained in their highest condition for the performance of their various functions.

Shoeing.—A good deal of controversy has been indulged in as to the best methods of shoeing and the form of the shoe, which it is hardly necessary for us to enter upon, beyond pointing out the necessity there is for employing a farrier possessed of sufficient knowledge and observation to do his work properly, as shoeing is often an operation requiring considerable delicacy and judgment, and should never be confided to a man who is not thoroughly master of his business.

Though the process of shoeing is generally gone through without any difficulty in common cases, through the docility of the animals, yet accidents have frequently happened to both men and horses from the violence of the former and the unsteadiness of the latter.

Lameness is frequently caused in the process of shoeing by what is termed a prick—*i. e.*, one or more nails enter the sensible parts within the hoof, or approach it so nearly as to cause pain when the weight of the horse's body resting on the foot forces it against the tender part.

When this tender part has been pricked lameness quickly follows the injury and suppuration succeeds, which sometimes runs under the sole and causes considerable derangement. When the quick is not penetrated it is often some days before the lameness becomes visible, and in which leg it is actually situated is often a matter of doubt. The pain which the horse feels in the injured leg causes him to step short with the sound one, so as to take the weight off the lame leg, which often induces people to believe he is lame in the wrong leg. The horse puts out the lame limb quickly and drops upon the sound one.

In these cases the shoe should be taken off and the seat of injury discovered by the tenderness evinced upon pressure. The foot should be pared out and the injury exposed, so that any matter that may have formed can escape freely, or if not already produced, that it should not be confined when it afterwards forms. The foot should be well fomented and afterwards placed in a linseed poultice, which should be continued for several days, unless the case is slight. The wound may need to be stimulated with a little tincture of myrrh, or, should granulations make their appearance, caustic should be applied, such as the muriate of antimony. The same mode of treatment should be adopted for bruises of the foot, which may occur from pressure of the shoe on the sole from being left too flat or not seated.

In cases of corns the shoe will require to be put on so that no pressure comes against the affected part, for which purpose the ground surface of the shoe needs to be bevelled or seated off for the space of an inch or more, so that when the foot touches the ground the heel receives but little pressure. This is assisted by leaving a space of an eighth or sixteenth of an inch between the heel of the foot and that of the shoe. When the feet are weak and flat it may be necessary to apply a round or bar shoe, by which means the heel can very effectually be secured from pressure.

In cases of sand-crack a bar shoe should be put on, so as to take the bearing off the affected part, and when the shoe is put on the hind foot precaution must be used that it does not touch the horn for the space of an inch at least on each side of the toe.

BREEDING.

The mare goes with young eleven months, and sometimes for nearly twelve, the greater number of foals being dropped at the end of May, which is, doubtless, the occasion for the age of all horses being reckoned from the 1st of May—except thorough-bred horses, whose years are dated from the 1st of January.

It is seldom, however, that they are foaled so early as January, though February is a very common time, the object of the breeder for the turf being to get his foals dropped as early in the year as he can, for the very sufficient reason that, being required to run at three years, and sometimes even at two years old, even a single month will give an advantage in age and strength. This plan, however, would not be suitable with other kinds of horses, as mild weather is desirable as well as a good bite of grass at the time of foaling.

In breeding, it is necessary to select parents possessing those qualifications which it is sought to be obtained in the offspring. Faults, as well as good qualities, are perpetuated, and it is idle to expect valuable progeny from parents which are defective in many particulars, and, of course, the best mares on the farm should be selected for the purpose of breeding—though sometimes one which has met with an accident, such as that of having broken knees, would answer as well to breed from as a sound mare, provided her constitution was perfect and there were no other defects.

In selecting a stallion with a view of breeding draught horses, a large compact horse should be chosen. A symmetrically formed horse mostly appears smaller than he really is, which often deceives a bystander as to his correct height, and it is a good point for him to be comparatively short in the legs.

Large cart horses are very much disposed to have ring-bones and side-bones on the pasterns, and a horse lame from this cause should be rejected. The hock is a very important joint, being much called upon in heavy draughts, and consequently liable to strains; and all affections of this joint, whether curbs, spavins, or thoroughpins, should be a reason for the horse being rejected.

The hocks should be broad in front, and neither too straight nor too crooked, nor yet cathammed, as in the case of shaft horses at times when a heavy load devolves upon them, when in the action of walking the weight they have to support is thrown alternately upon each hock.

The eyes of an old horse should be free from all appearance of disease, unless from accident, which will be immaterial; while in the case of a young one, there should be not only no actual disease, but the organ should be free from any tendency to it.

They should be full without being too convex, the small sunken eye being much more liable to disease than the large clear eye.

As cart horses have a greater predisposition to swellings and humours than saddle and other horses, it is necessary to guard against this evil by choosing a stallion as free as possible from this inclination, and the fore legs should be strong and flat below the knee, and not round and gummy either before or behind. The forearm should be strong and muscular, and should not stand too much under the body, while the shoulders should be tolerably oblique, as when the shoulders are good there is a fair presumption that the horse is likely to be a good walker. The elbow should not be too close to the chest, but there should be room enough to put the hand between them. The neck will be better too thick than too thin, of average length, and, if moderately arched, it will be advantageous, for it is a great fault in any horse to have an ewe neck. The angles formed by the juncture of the neck with the body, and by the head with the neck, should not be too acute, for such horses are liable to *poll evil*, from the disposition induced of throwing up their heads suddenly, and striking their poll violently against a low doorway or other object above them. The back should be straight and broad, with the ribs well arched ; and the false ribs of due length, so as to give the abdomen capacity and roundness. The quarters should be full and muscular, and the tail well set out and not too drooping, and the horse should girth well.

The tendency of many large horses is to have thin horn and flat feet ; and the feet being a point of great importance, a stallion possessing such feet should be looked upon as objectionable. The feet had better be too large than too small, and plenty of horn is a desirable feature.

The mare should be free from vice and vicious habits, and possess as many of the good points which have been enumerated for the stallion as possible.

A good horse is always of good colour, according to the old saying, yet white legs and feet are objectionable. It is well known that white hoofs are more tender and somewhat more liable to accidents than black ones, and horses when thus marked, do not fetch so much money, while in most instances at exhibitions they are debarred as prize takers.

When once possessed of a good stock, it may easily be kept up by breeding from a couple or more of the mares at a time, by which very little interruption will be occasioned to their work, and thus almost insensibly, without any apparent outlay, teams are maintained which otherwise would cost large sums to renew.

Parturition.—The young of the mare is generally brought forth without much inconvenience, and it is seldom that she requires assistance, but in difficult cases the plan adopted is to insert the hand or arm, and push back the fœtus, and bring forward the parts which ought to be presented first.

Foals.—The management and treatment of young cart stock is so little under the control of any general system, and is so commonly understood, that little more is required beside an abundance of wholesome food and care to guard them from exposure to the weather in the winter months, and prevent them from lying in the wet. Good day hovels are better as nightly shelter for foals than warm stables, and colts thus treated will generally have acquired sufficient strength and hardihood when two years old off, as we have stated before, to be put gently to the plough in the following spring.

The breeding of race horses and pedigree horses being made a special business of itself, an account of the various methods followed will scarcely fall within the compass of the present work.

CHAPTER VII.

HORSES (Continued.)

Diseases and their Treatment—Apoplexy, Megrims, or Vertigo— Inflammation of the Bladder, or Cystitis—Spasm of the Bladder—Inflammation of the Bowels, or Enteritis—Bronchitis— Calculi, or Stones in the Intestines—Canker—Catarrh, or Cold —Chopped Heels—Colic, Gripes, Fret—Chronic Cough— Diabetes — Diarrhœa— Dysentery—Farcy—Glanders—Grease —Diseases of the Heart—Hernia—Influenza—Strangulation of the Intestines, and Rupture of the Intestines—Jaundice— Inflammation of the Kidneys, or Nephrites — Laminitis, Founder, or Fever in the Feet—Inflammation of the Liver, or Hepatitis—Mange—Navicular Disease—Palsy (Paralysis) Pleurisy—Pleuro - Pneumonia—Pneumonia—Pumiced Feet— Rabies, or Canine Madness—Rheumatism—Roaring—Staggers-Mad, or Inflammation of the Brain—Stomach-staggers —Spavin—Splint—Tetanus—Thrush—Broken Wind—Thick Wind — Worms.

Diseases and their Treatment.

Apoplexy, Megrims, or Vertigo.—This is caused by sudden determination of blood to the head, sometimes produced by a tight collar, and induced by high feeding, particularly in the spring or early part of the summer. It is dangerous to use horses predisposed to this disease, and it increases with age. The horse suddenly stops in his work, shakes his head, and often falls ; at others, giddiness only is manifested, and after a while the horse rallies and moves on.

Bladder, Inflammation of the, or Cystitis.—A painful disease, not of very frequent occurrence, often produced by a chill, and from irritating subjects received into the bladder. It is evidenced by continual efforts to stale, when the bladder is emptied of its contents, and can contain no urine. Bleeding should be resorted to, and a sedative ball given as advised for inflammation of the kidneys, and the bowels opened by means of linseed oil. If a mare, thin linseed tea should be injected into the bladder.

Bladder, Spasm of the.—This, or more properly, spasm of the neck of the bladder, arises from the animal not being allowed to stale for a long period. The bladder being distended, gets inflamed, and will even burst. The treatment consists of the insertion of a catheter into the bladder to pass off the urine, which in the case of the mare is easy, but difficult in that of the horse. If there is much irritation bleeding is resorted to, and an injection and opiate administered.

Bowels, Inflammation of the, or Enteritis, is rapid and violent in its progress, and the animal mostly lies down; the pulse quick, small and thready, and the membrane of the eyelids and nostrils are intensely injected. It is commonly occasioned by exposure to wet when in a heated state, such as passing through a stream of water, or drinking copiously of cold water; and by indigestion and unwholesome food.

The best treatment is to bleed freely, and if the pulse is weak previous to bleeding, an ounce of spirit of nitrous ether should be given with the same of laudanum, after which the horse will bleed better. Draughts of linseed oil should follow, and a solution of opium given every four hours. Hot applications and frictions to the extremities are also recommended.

Bronchitis is inflammation of the mucous membrane lining the bronchial passages, or air tubes of the lungs, and is the more dangerous disease from its insidious nature, for, commencing as a common cold, it all at once assumes dangerous symptoms, which not unfrequently terminate fatally.

Its presence is indicated by a quick, and often a very weak pulse, with short and weak cough, which seems to proceed from the chest, and causes pain. The discharge from the nostrils is copious, sometimes tinged with blood, or of a brown colour. The extremities, though occasionally hot and cold alternately, are mostly warm; there is an accelerated respiration and an indisposition to lie down. As it is not considered safe treatment to administer laxative medicines, moderate bleeding, with the finger on the pulse, is recommended.

The same ball may be given as for catarrh, as well as stimulating the throat and blistering the course of the windpipe. A seton also inserted in the brisket is sometimes advisable. Counter irritation should be actively resorted to in severe cases. Good nursing is imperatively called for, and the subject should be fed upon linseed and oatmeal gruel, with grass in summer and carrots in winter.

When the inflammatory symptoms have been got under, a mild tonic ball composed of the following will be found efficacious :

Gentian (powdered)	2dr.
Pimento (powdered)	1dr.
Sulphate of iron	1dr.

made up into a ball with treacle, and given once a day. Occasionally it will be prudent to omit the iron.

Calculi, or Stones in the Intestines, are sometimes produced by the horse eating some foreign substance, which forms a nucleus round which portions of food accumulate, which give rise to the symptoms of colic. When not very large they are passed off with the dung, but in serious cases become impacted in some portion of the intestines.

Care in feeding is the best preventive of this disease, which is a very difficult one to treat, relief only being afforded by the dislodgment of the stone.

Canker is of a similar nature to thrush, but worse, often extending over the whole foot, and sometimes being incurable, commonly arising from an injury to the foot from a nail, bruise, or from grease, or neglected thrush. A strong caustic should be applied to the granulating surface with a view of stopping the discharge and diseased growth, and promote the secretion of horn, such as hydrochloric acid, or, if the growth is considerable, the knife will soonest make a sound surface, the bleeding being stopped by the hot iron. The secretion of horn is promoted by the application of tar, and sulphate of zinc becomes, after a time, a useful application. Dirt and moisture must be avoided,

K

and care must be taken not to irritate the surface too much by violent treatment.

Catarrh, or Cold, is inflammation of the membrane which lines the chambers of the nostrils and the throat. When the latter is affected it is called sore throat. Sneezing comes on first, which is followed by a cough and a discharge of mucus from the nose. The symptoms may be either slight or severe, but when, in the case of the latter, it leads on to bronchitis, it is attended with danger.

Change of temperature from heat to cold, or *vice versâ*, is the most fruitful source, some horses being much more subject to colds than others, young horses being more liable to catch them than old ones. A few bran mashes will dispose of slight cases without any regular medical treatment, but in severe cases, with a strong pulse, bleeding is advised, and the throat should be stimulated externally with tincture of cantharides. Should the bowels be costive, two or three drachms of aloes should be given. If not, the following cough ball, administered night and morning for several days :

Nitrate of potash ... ... 2dr.
Tartarised antimony ... ... 1dr.
Digitalis (powdered) ... ... 1sc.
Linseed meal ... ... 8dr.

mixed up into a ball with Barbadoes tar.

Bran mashes and soft food should be given, and in bad cases oatmeal in linseed gruel.

Chopped Heels is a breach in the continuity of the skin, arising from wet and cold to the legs, which inflames the skin and parts beneath. Purgatives should be given, and linseed poultices placed on the heels, and exercise avoided. After two days the following lotion should be applied to the heels :

Sulphate of zinc ... ... 4dr.
Alum (powdered) ... ... 4dr.
Water ... ... 1pt.

After the physic has ceased to operate the following powder may be applied daily to the chap :

Powdered chalk ... ... 1oz.
Sulphate of zinc ... ... 1dr.
Alum ... ... 1dr.
Bola Armenian ... ... 4dr.

Choking.—An obstructing object should be first endeavoured to be removed by the hand, but if too far down a flexible probang should be used, first being well oiled. If these fail the œsophagus must be carefully opened, and the obstruction removed, and the wound skilfully sewn up. The horse must be kept without solid food for several days afterwards.

Colic, Gripes, Fret.—There is *flatulent* colic, the distention of the stomach or bowels with gas; *spasmodic* colic, or violent contraction of the coats of the intestines; and *stercoral* colic, or distention of the intestines with food. Indigestion, unwholesome food, and succulent grasses chiefly cause flatulent colic; while the spasmodic form often proceeds from a sudden draught of cold water; and unwholesome food and ravenous feeding produces the stercoral form.

A draught containing an ounce of tincture of opium, with 2oz. of spirits of nitrous ether, will sometimes afford immediate relief. An ounce of sulphuric ether will be better than the nitrous ether in the flatulent variety, condensing the gases more; to this an ounce of tincture of aloes or valerian may be added. The horse should be bled rather copiously, and another draught administered if relief does not follow, after which oily purgatives, such as linseed oil; 1lb. combined with smaller doses of opium, and half a drachm each of calomel and tartarised antimony, may be given every four hours, till three or four doses are taken. In stercoral colic, the last method of treatment should be resorted to first. In obstinate cases, frictions and hot fomentations to the abdomen should follow.

Cough, Chronic, arises from irritability and too great dryness of the air passages, and is a frequent attendant on thick wind. It may be occasional or frequent, and on the accession of a fresh cold the old one becomes increased. When the cough becomes worse the throat should be stimulated, and the cough ball suggested for catarrh given. In bad cases a seton under the throat has done good service.

Diabetes, or excessive staling, is very troublesome and often difficult to cure, the urine inclining to form sugar, which is produced by unwholesome food, such as mowburnt hay, or kiln-dried

oats. Wholesome food must be substituted, and the following
ball given twice a day :

Sulphate of iron	1½dr.
Gentian	2dr.
Ginger	1dr.
Opium	½dr.

with treacle to form a ball. Water should be allowed without
stint, but thin linseed tea is better.

Diarrhœa arises from the irritated state of the mucous membrane lining the intestines. New oats and hay, as well as fresh
grass and other green food, have a tendency to produce this
derangement.

Food of a more wholesome and binding nature should be given,
such as beans, and if medicine is necessary, the following in thick
gruel made with flour may be administered :

Ginger (powdered)	1dr.
Gentian (powdered)	2dr.
Opium	½dr.
Prepared chalk	1oz.

repeated two or three times a day.

Dysentery also consists of an irritated condition of the mucous
membrane lining the intestines. The natural mucus secretion is
greatly increased, so that the dung is passed encased in mucus,
in a half solid form. The fæces may either be relaxed, or in a
very indurated state, but in either there is a copious secretion of
mucus, and often an offensive smell. The disorder varies in
intensity from a mild to a dangerous form, being often associated
with other diseases, such as influenza.

Mild bleeding should be resorted to, and oily laxatives given,
with linseed gruel in tea. Two drachms of nitrate of potash
and four drachms of super-tartrate of potash may be given with
gruel several times a day, and bran mashes and carrots instead
of the ordinary food.

Farcy is a different manifestation of the same disorder as
glanders ; unventilated stables and bad provender, coupled with
hard work, being the most likely origin of the disorder. The
first symptom is generally lameness, with swelling of one of the
hind legs, on which a wound appears. If the disease is confined

to a single limb, it may, at times, be treated successfully, the following ball being given twice a day :

Sulphate of iron	1dr.
Gentian (powdered)	1½dr.
Pimento (powdered)	½dr.
Iodide of potash	5gr.
Cascarilla bark	1½dr.

made into a ball with treacle. The hair should be cut from the enlarged absorbents, and either a tincture of iodine well rubbed in, or a mercurial linament with iodine ointment.

Glanders.—The ravages of this disease amongst coach and waggon horses was at one time very formidable, but is now by no means so common. It consists of a discharge of postulent matter from one or both nostrils, with a hard enlargement of the sub-maxillary glands. The cures of this disease have been so very rare that it is by no means profitable or desirable to make the attempt.

Grease is an offensive discharge from the heels, which often follows chapped heels, and, when neglected, becomes chronic. The treatment should be the same as for chapped heels, but pushed on vigorously and continued longer. Lameness, corns, and sand-crack we have briefly referred to under the head of shoeing. False-quarter is often the result of neglected sand-crack.

Heart, Diseases of the.—The heart of the horse is not very liable to disease, carditis, or inflammation of the heart, being rare, and generally mixed up with other diseases ; peri-carditis, or inflammation of the membrane which invests the heart, being more common, and often mixed up with pleurisy, which it very much resembles, and is indicated by laboured action of the organ, and hypertrophy, or enlargement of the sides of the heart ; but the symptoms are generally obscure, and the treatment unsatisfactory.

Cancer of the heart is also sometimes found, the heart being enlarged and mis-shapen. Rupture of the heart is somewhat uncommon, but spasm of the heart is more frequently met with, being often produced by over exertion, when the beating of the heart can be felt all over the body. If the animal is put to fast work before the heart recovers its former tone, a fatal result will take place. The following draught should be given at once :

Spirits of nitrous ether	2oz.
Tincture of opium	1oz.
Tepid water	12oz.

and repeated if necessary; and the bowels should be afterwards
regulated by a mild laxative and bran mashes; several days' rest
being allowed.

Hernia is not very common in England, where horses are
mostly castrated, entire horses being most subject to it; a portion
of the intestine escapes into the abdomen, and passes into the
scrotum with the spermatic cord.   The best treatment to adopt
is what is called the covered operation of castration, by removing
the testacles without cutting into the cavity in which they are
contained, in the case of colts when scrotal-hernia is congenital, or
formed at birth.   Great caution as well as skill are demanded, and
the operation should only be attempted by the most skilful hands.

Influenza is a kind of low nervous fever, attended with great
prostration of strength, affecting more particularly the mucous
membranes.   In other instances it appears to connect itself with
severe inflammation of the viscera of the chest and abdomen.

A diffusive stimulant, as the following, should be first given:

Spirit of nitric ether	1oz.
Potassio-tartrate of antimony	1dr.
Nitrate of potash	4dr,
Warm water	10oz.

to be repeated if required.

After the draught has been administered six hours, the following ball may be given twice a day:

Proto-chloride of mercury	2sc.
Potassio-tartrate of antimony...	2sc.
Nitrate of potash	2dr.
Linseed meal	3dr.

made into a ball with soft soap.

After fever has subsided, and debility and want of appetite
remain, the following tonic may be given twice a day:

Gentian (powdered)	1½dr.
Pimento (powdered)	½dr.
Sulphate of iron	½dr.
Linseed meal	2dr.

made into a ball with treacle.

Intestines, Strangulation of the, and Rupture of the Intestines.—
The cause of these is generally obscure, sudden exertion with an
overloaded stomach doubtless being the most frequent cause.
The treatment of these diseases is quite beyond the power of any
but a veterinary surgeon.

Jaundice consists of absorption of the bile, which, entering the blood, tinges all the membranes a yellow colour. It is, however, not a common disease, nor a very serious one, a derangement of the appetite being the chief symptom.

The ball recommended for hepatitis will be the best treatment, but it may be preceded by three or four drachms of aloes.

Kidneys, Inflammation of the, or Nephrites, is not very common, but is generally acquired from a cold chill across the loins, or sometimes a strain or injury may produce it. Great pain when the loins are pressed, the urine dark in colour, and sometimes black, accompanied with fever, quick pulse, and disturbed respiraation, are the indications.

Copious blood-letting must be resorted to, followed by a purgative draught, with frequent injections. A fresh sheep skin should be applied to the loins, and renewed in the course of twelve hours, and mustard poultices applied. A sedative should be given twice a day as follows :

Tartarised antimony	1dr.
Opium	½dr.
Proto-chloride of mercury	1dr.
White hellaboro	1sc.

made up into a ball.

Laminites, Founder, or Fever in the Feet, consists of inflammation of the sensible laminæ which connects the coffin bone with the crust, and which mainly secretes the latter. It is produced by long-continued standing and prolonged violent exertion, sometimes following inflammation of the chest or pleurisy. The feet are hot and tender, and if the animal has to move up when lying down he treads on his heels with an exhibition of pain, and can scarcely be made to stand. The shoes must be taken off and a large quantity of blood extracted from the feet or the arms. Setons may be inserted in the frogs and linseed poultices applied to the feet. After a while, blistering the coronet will be useful.

Liver, Inflammation of the, or Hepatitis.—The presence of this disease is evidenced by loss of appetite, short and rapid breathing, and a yellowness of the eyelids and nostrils, and the dung is usually hard and coated. The disease is longer in its duration than inflammation of the lungs, and the symptoms are not so acute.

High living and want of exercise are the causes in most cases, and it must be met by bleeding in the first instance, to be followed by a blister on the right side, and small doses of calomel and opium, the bowels, if in a costive condition, having been first opened with linseed oil. The following ball should succeed, given twice a day:

Calomel	1dr.
Opium	1sc.
Nitrate of potash	2dr.

This ball should be given for several days.

Mange.—Mange is an offensive disease, unfortunately only too common, and is highly contagious, communicated even by the hands of an attendant, or contact with harness. The disease being in the skin this must first be cleaned thoroughly, after which the following linament should be thoroughly rubbed in:

Sulphur vivum	4oz.
White hellebore	2dr.
Oil of tar	4oz.
Linseed oil	1lb.

well mixed together.

Navicular Disease is common in light, well bred animals, but scarcely ever occurs in the heavy cart horse, and consists, in the first instance, of inflammation of the membrane which covers the cartilage of the navicular bone, as well as the tendon which corresponds to the bone, the inflammation being followed by ulceration. Hard roads and the fast pace, alternated by long confinement, to which may be attributed the hard and dry foot which generally accompanies this disease, are the most frequent causes, and it is only in the early stages that a cure can be hoped for. Blood should be taken from the foot to the extent of 8lb., or, if both feet are affected, 6lb. may be taken from each foot. The sole should be pared thin, and the quarters well rasped, and the feet put in a linseed meal poultice for a week, after which a seton should be inserted in the heel and kept there for a month. The pasterns may also be blistered.

Palsy (Paralysis).—Generally produced by a slip or fall, by which the spinal cord becomes injured, so that the horse is unable to stand, and can only be supported by means of slings. Treatment is rarely of much service, and when partial recovery takes place, the horse becomes what is termed *chinked* in the back, and incapable of sustaining burdens. Bleeding and blistering the spine are the principal remedies.

Pleurisy, or inflammation of the membrane which lines the interior of the chest as well as the lungs and its other contents, arises from exposure of the body to cold when heated and from sudden alternations of temperature. The horse will lie down, which he will not do in pneumonia. Severe pain is at times exhibited, and there is tenderness on pressing the ribs.

Active blood-letting is prescribed till the pulse becomes almost imperceptible, and repeated, if required, once or twice within from six to twenty-four hours. The sides should be blistered and the action of the blisters kept up. The same ball should be used as prescribed previously.

Pleuro-Pneumonia is a compound of pleurisy with pneumonia, and is a fatal disease, somewhat obscure, occasionally assuming an epizootic form, destroying life under the name of influenza. With oxen a disease of this nature has been dreadfully fatal. The treatment should be modified and conducted in accordance with the principles laid down in the following. Blood-letting is more doubtful in its results than in diseases of a more defined character.

Pneumonia.—There are two forms of inflammation of the substance of the lungs, congestive and ordinary pneumonia, which are brought on by over exertion, or from alterations from heat to cold, and the reverse.

The congestive form is indicated by a very quick and weak pulse, rapid and distressed breathing, and cold extremities. Very prompt treatment is necessary, as death sometimes follows in four-and-twenty hours. The progress of the disease is less rapid in ordinary pneumonia, taking three, four, or five days, as the case may be, before a fatal termination ensues. The air cells become filled with lymph, and can no longer perform their functions; and suffocation finally takes place. The breathing is greatly accelerated as well as the pulse, which may be either strong or weak; the extremities are cold, though not so much so as in congestive inflammation, and the animal remains in a standing position.

Bleeding is called for, but if the pulse is weak, it is best to give the following in half a pint of warm water:

> Spirits of nitric ether ... ... 2oz.
> Solution of acetate of ammonia ... ... 1oz.

This will bring warmth to the skin and cause the animal to bleed better than otherwise; the amount of blood taken to be regulated by the strength of the pulse, extending from six to twelve pounds, and may be repeated if necessary in the course of twelve or twenty-four hours. The sides should be blistered and a seton placed in the brisket.

Every six hours a ball composed of the following should be given:

> Nitrate of potash ... ... 2dr.
> Proto-chloride of mercury... ... ... 0½dr.
> Tartarised antimony ... ... 1dr.

When six of these balls have been given, those recommended for catarrh should be given less frequently. After 2oz. of the nitrate of potash have been taken it should be discontinued.

Pumiced Feet.—This term is applied when the soles become convex instead of concave, and the crust is increased in obliquity, uneven and furrowed. The disease is principally confined to heavy horses, where the strength of the crust is not equal to the weight of the horse, and thus the foot gradually sinks. No cure can be effected, but relief can be afforded by shoeing, so as to prevent the sole from receiving pressure, and yet protect it from injury.

Rabies, or Canine Madness.—Generally caused by the bite of a rabid dog, the poison of which will remain in the system from six weeks to three months, and occasionally longer.

No treatment is successful when the symptoms have manifested themselves, but the disease may be prevented if the bitten part is carefully cauterised soon after the injury.

Rheumatism.—Rheumatism in the horse generally takes an acute form, and is then designated a chill, the muscles and tendons being the seats of the disease, and it sometimes precedes or follows pleurisy.

Bleeding is recommended, after which the bowels, being usually costive, should be opened by aperients. Six drachms of aloes, with two of ginger, may be dissolved in hot water, and given with one or two ounces of spirit of nitric ether, after which the following ball may be given twice a day :

Proto-chloride of mercury...	2 sc.
Potassio-tartrate of antimony...	2 sc.
Nitrate of potash	2 dr.
White hellabore	1 sc.
Linseed meal	4 dr.

to be made up into a ball with soft soap.

Roaring is a disease of the larynx or windpipe, when there is a partial obstruction of air to and from the lungs. As this may arise from several causes, it is not always easy to find out by what it is occasioned, and the symptoms must be closely watched so as to draw a fair inference of the cause. If it succeeds an attack of catarrh, or is accompanied by a cough, blisters or setons will afford relief.

Staggers-Mad, or Inflammation of the Brain.—Phrenitis is more common with heavy than light horses, and arises from fulness of the blood vessels, from too high feeding and want of exercise. Profuse bleeding should be resorted to, either from the jugular veins or carotid arteries ; six or eight quarts of blood may be taken, or until the pulse can scarcely be felt. The horse should be well purged, and fever medicines given after, with cold applications to the head.

Stomach Staggers, arises from distension of the stomach with too much food, but is rare where there is good management in the feeding of horses. The brain, in sympathy with the stomach, becomes oppressed, and the horse appears sleepy, and forces his head against any object before him.

Ravenous feeding when the stomach is empty, or the food is dry and difficult of digestion, produces it ; commonly occasioned formerly when heavy horses were allowed to fast many hours, which the invariable use of the nose-bag is now often the means of preventing.

Oily purgatives should be given, assisted by draughts of warm water, and injections containing a purgative. Stomachics, such

as carbonate of ammonia 2dr., gentian 1dr., with 1oz. of spirits
of nitrous ether, twice a day, may be given, The disease is often
fatal, the stomach being distended at times beyond the power of
contraction.

Spavin.—Spavin often causes permanent pain and lameness,
being a more serious ossification than splint, situated near a most
important joint, on which the whole weight of the horse is
thrown, the disease often spreading on the articulating surface of
the joints, somewhat after the nature of navicular disease. The
higher the spavin is situated, the more likely to occasion lameness.

The internal disease referred to causes treatment to be unsatisfactory very often, but the proper course is either to use the
firing-iron, to blister, or apply a seton.

Fractions, dislocations, and wounds mostly require the services
of a veterinary surgeon, as indeed do the majority of the serious
diseases we have mentioned, and it will always be found the best
course to call in an experienced man when occasion demands,
although at the same time it is necessary that the owners of
valuable animals should be made acquainted with the outlines
and forms of the most common diseases.

Splint is a common disease, situated between the large and
small metacarpal bones, and generally on the inside. The bones
in the young animal are connected together by ligamentous substances, which become stretched and inflamed, and the vessels,
thus excited, throw out a bony deposit, which, being deposited
under the periosteum, or covering of the bone, puts it painfully
on the stretch and causes lameness.

A small incision is made through the skin at the upper and
lower part of the splint, and passing up a small narrow knife and
so cutting in and dividing the periosteum, which thus relieves
the tension and the irritation. A small seton from one incision
to the other keeps up a counter irritation for two or three weeks.
In slight cases a blister may be applied, and the following will
be found efficacious:

Iodide of mercury	1dr.
Palm oil	1oz.

well mixed.

Tetanus.—This is frequently termed locked-jaw, though the latter is but one of its symptoms, and when confined to the neck is called *Trismus*, and is more manageable than when the greater part of the body is convulsed. It is brought on by local injury, such as the wound of a joint, or sometimes a slight cut. Exposure to cold and wet will produce it, and sometimes its origin cannot be traced.

Copious blood-letting should be resorted to, and powerful purgatives given, as aloes, eight drachms, or forty to sixty drops of croton oil, succeeded by opium and camphor, in doses of a drachm each. Blisters to the abdomen have been found useful, and the animal should be kept as quiet as possible. Gruel should be given to drink, administered between the teeth, if possible, or by the stomach pump.

Thrush is an offensive discharge from the cleft of the frog, which rots it away, and causes the horse to be hurt when treading upon a stone. It is produced by moisture and filth, yet sometimes engendered in the forefeet by contraction and heat. Thrush seldom causes permanent lameness, but is the occasion of many a fall.

The cleft being well cleaned out by a pledget of tow being drawn backwards and forwards, should be dressed with the following ointment:

Oil of turpentine ... ... 4dr.
Sulphuric acid ... ... 4dr.

gradually mixed together in an open place, and after the boiling has subsided which takes place, add to it:

Barbadoes tar ... ... 8oz.
Palm oil ... ... 4oz.

The same ointment will be found good for corns, and assist in promoting the growth of healthy horn after it has been pared away at the seat of disease.

Wind, Broken, consists of a rupture of the air cells of the lungs, so that the air escapes from them and inflates the pleura which covers the lungs. Though a double expiration takes place, as it were, yet there is not the same capacity for inhaling fresh air, and the animal is, consequently, incapable of performing the same amount of rapid exertion. In the case of farm horses, the somewhat coarse, bulky, and dusty food they get aggravates the disorder. Sudden exertion on a full stomach is the cause, while foul and dusty provender may be regarded as the remote agent in producing the disease.

The only treatment which can be followed is that of palliation, and the horse kept in the highest condition to enable him to accomplish ordinary work with as little muscular exertion as possible, so as to make less demand on the lungs. The food should be of a concentrated kind, and the stomach never distended—very little hay or straw, but carrots given instead. If medicine is necessary, the cough-ball prescribed for catarrh will be efficacious. Water should only be given in small quantities, and the horse should not be called upon for much exercise with a loaded stomach.

Wind, Thick, is most commonly caused by chronic attacks of inflammation of the lungs, and the same treatment as in broken wind will be desirable.

Worms.—Horses are subject to the visitation of several kinds of worms, the most common of which is a short thick grub called the "bot," which is sometimes found in large numbers attached to the internal surface of the stomach by a kind of hook. Towards the spring of the year they pass off with the dung, and are hatched by the action of the sun into a species of gadfly, which secretes its eggs on the skin of horses, the irritation causing the animal to bite the part, when the eggs are swallowed, and are hatched in the stomach.

The *long round* worm is also frequently found in the intestines of horses, but not being very numerous, they do not appear to exert any very injurious effect upon the health of the horse, the most common as well as the most injurious being small thread-like worms, called *ascarides,* usually found in the large intestines, and particularly the rectum, where they sometimes cause great irritation.

The presence of worms may be inferred when, in spite of good food and proper care, the horse appears out of condition, and his coat is harsh and rough; the true state of affairs may be surmised if worms are found in the dung. When *ascarides* are present, there is usually some irritation manifested at the anus, and an ejection of a white secretion from this part.

The presence of a worm or two in the dung, particularly of the long white sort, at rare intervals, is a matter of little moment, unless there is plainly an appearance of a want of condition about the horse ; but if the worms are numerous he should have bran mashes for a day or two, and the annexed administered :

Tartarised antimony	2dr.
Spirits of turpentine	3oz.
Linseed oil	1½lb.

well mixed together. After this the following ball may be given daily for a week :

Sulphide of iron	1dr.
Powdered gentian	2dr.
Powdered ginger	1dr.
Powdered pimento	1dr.

made into a ball with treacle, which will materially help in restoring lost condition.

To those who desire further information on the diseases of horses, written untechnically and intelligibly, we recommend Mr. Dalziel's cheap little handbook on the subject, issued by the publisher of this book.

CHAPTER VIII.

COWS.

Various Breeds of Cows—The Shorthorn—The Alderney—The Ayrshire—Kyloes, or West Highland—The Galloway—The Welsh—The Irish—The Hereford—Longhorns—North Devons —The Suffolk Dun—Breeding—The Economical Rearing of Calves—The Feeding of Cows—Attention to Stock.

VARIOUS BREEDS OF COWS.

Cows are the most profitable of all stock, there being always a ready sale for produce in the form of either milk, butter, or cheese; but as the operations of the dairy form a large subject of itself, requiring special reference to different forms of management, an explanation of the various processes pursued form no part of the scope of the present little work, which is devoted to the consideration of stock-keeping as a branch of husbandry by itself, outside that of dairy farming, which, however, is a department that ought to receive more attention than it does from the ordinary arable farmer. It is often assumed by the latter that, not having any pasture land of consequence, he is not in a position to turn his attention to dairy farming successfully, overlooking the fact that, with only a limited area upon which to turn out cows for air and exercise, yet with only slender means in this way, a certain number of animals could be very profitably kept upon the soiling system, upon the plan pursued in some of the large London dairies when the animals are sold off quickly and the stock frequently renewed.

When this system is carried out, the large breeds of cattle such as the shorthorn, are usually made use of, a race of animals that put on flesh quickly when fattening, and by the flavour of their meat after this process has been properly carried out, show no signs of having been used for dairying purposes, and we will commence our description of the various breeds of cows with this important class.

The Shorthorns.—The shorthorn has arrived at its present pitch of excellence by means of the careful and skilful breeding which it has received, and it has now well earned its title, "Improved Shorthorn." The shorthorns are distinguished over all other cattle by their aptitude to fatten, which enables them to attain a maturity and weight of carcase at a very early age, which no other race can approach. They possess symmetrical forms, and a quiet even temper, and in colour vary from pure white to a rich red, and often a mixture, sometimes forming a deep or light roan, occasionally termed hazel or strawberry. The excellence of the breed has caused them to be in request not only in every European country but also in America and our colonies, where the pure breed is now as solicitously kept up as in England, where individual animals of the best pedigrees have realised sums at different stock sales ranging from £200 to £700 per head. The cow partakes of the general characteristics of the male, except that her head is finer, longer, and more tapering than that of the bull, and her shoulders are more inclined to be narrow towards the chine.

As grazing cows, on account of their large size, they require rich pastures, and are not well adapted for poor ones, while their aptitude to fatten and lay on flesh, rather than secrete milk, does not allow them to be such profitable dairy cows in proportion to the cost of the food they consume, as many of the breeds we shall subsequently mention, though they are very suitable for the purpose of the arable farmer, who feeds them upon the hand system, and can cut and bring to them an abundance of rich, succulent food, for which they are admirably adapted. The London dairymen stimulate the production of milk by feeding with brewers' grains and other milk-making articles of food, and pursue a definite system of management, which makes this breed a very valuable one to them.

Instances are on record of a very large yield of milk being

L

given by different shorthorn cows, but, notwithstanding all that may be said to the contrary, their readiness to fatten, which is their most distinguishing trait, is naturally opposed to the secretion of milk, which is the best quality that a dairy cow can be possessed of.

The Alderney.—The Alderney gives but a comparatively small supply of milk, but it is rich in cream, and consequently enjoys a high reputation as a butter cow, as much, or perhaps more, butter being obtainable from her produce than from that of a cow which gives double the quantity of milk. On this account the Alderney has always been a favourite breed with private gentlemen, who wish for a superior quality of dairy produce. Alderneys are content, and will do well upon more inferior herbage than would satisfy the wants of a grazing shorthorn.

The Alderney is not a breed that will answer the purpose of the stock-keeper who wants to do a trade in milk. But upon somewhat poorish land (though it must not be too poor, for which another breed we shall next mention is better fitted), in an inland situation, where there are no facilities for the disposal of milk by railway or otherwise, the Alderney will be found a useful animal ; or, if not exclusively kept, the richness of its milk will, in conjunction with that of cows of other breeds, improve the general quality of the butter obtained.

For the purpose of the grazier, to be fed as stock, the Alderney, as a breed, is totally useless, it being a work of some considerable difficulty to fatten them.

The Ayrshire.—Quite the opposite to the Alderney is the Ayrshire cow, which is remarkable for the abundance of its milk, which it possesses the power of producing from poor or medium herbage, and it has long been a favourite in many of the Scotch dairies, where great pains has been taken in breeding for the development of its milking powers. It is, therefore, an admirable dairy cow in those situations where the production of milk is aimed at, and where the feed at command is somewhat of an inferior quality.

Success in dairy farming mainly depends upon the consideration of points like these, and having a breed of cattle that is well adapted to the farm upon which they are to be placed ; for a small Ayrshire cow will do well, and bring satisfactory produce, in situations where the larger breeds would be scarcely able to support an existence.

This breed, alike with the Alderney, is not suitable for the purpose of the grazier, the beef being coarse in quality, and the animals can only be fattened with difficulty.

Kyloes, or West Highland.—We may as well continue our notice of Scotch cattle, and refer to the Kyloes, which is a race peculiarly well fitted to their native habitat, where the pasturage is coarse and scanty, and the climate humid, and it will be found a useful breed in similar situations, which are not adapted for any other breed of cattle, and hence is mostly given over to sheep farming. The cows give but very little milk, though it is of a rich quality, the value of the race chiefly lying with the capabilities of the ox and his feeding qualities, of which we shall speak again. As well as yielding only a small supply of milk, the cows have a tendency soon to get dry, which renders them undesirable as dairy cows, save in the exceptional circumstances we have referred to.

The Galloway.—Similar in its general characteristics to the Kyloe is the Galloway, which is a polled breed of larger frame than the former, and necessarily better fitted to somewhat richer pastures than its kindred race, possessing an aptitude to fatten, which causes them to be valuable as store stock, but as milch cows they have long been supplanted by the Ayrshire, and therefore need not be much taken into account.

Welsh Cattle.—There are two distinct breeds of Welsh cattle, the Pembrokeshire, which may be regarded as the type of the mountain races, a small hardy breed, something after the same standard as the West Highland or Kyloes, and are suited to a humid climate and coarse pasturage ; and a larger breed, which is found in the county of Glamorgan. Some excellent dairy cows are found amongst the latter, but the race is very much confined to the county from which it takes its name, while amongst the former are often found excellent milkers, some of them giving a good yield in proportion to their size, and the food they require.

There is also a heavier, coarser breed than the latter peculiar to the Anglesea district.

Irish Cows.—Injudicious breeding has caused the great majority of Irish cows to take but low rank as milk producers, the best being found amongst the small breed known as the Kerry, from the county in which they are chiefly reared, where some little attention has been bestowed upon them and an endeavour made to perpetuate good qualities. Of late, however, especially in the North of Ireland, more attention has been bestowed upon the breeding of stock, and it is probable that a better breed of cows will ultimately be found in the sister kingdom than is now generally met with.

The Hereford.—Hereford cows have long enjoyed a good reputation in their own county, being a large race of animals, especially well fitted for rich pastures and fertile soils, like those required for the shorthorn. Indeed, the lovers of this breed maintain that they are equal to, if not superior to, the shorthorn race; but as they have made but very little way out of their own county this conclusion is not universally adopted. Many of the cows give a good supply of milk, but they appear to be better fitted for the object of the grazier who has rich pastures to turn them out upon, and who desires to rear large animals that will come heavy to the scale for the butcher, than for the purpose of the ordinary dairy farmer.

Longhorns.—These are a similar race of animals to the preceding, and used universally to be met with in many English counties, receiving attention from the hands of that skilful breeder, Bakewell, whose name is so closely connected with the improved Leicester sheep, and whose excellent judgment in all matters relating to stock and breeding caused him to improve upon everything he touched, bringing conspicuously out the best points of every animal. The ubiquitous shorthorns have, however, displaced these, and they now no longer enjoy the reputation they once possessed.

North Devons.—These are a nice breed of animals, justly in favour on account of their excellent temper for working on the farms, when ox-teams were used extensively instead of horses.

They are not well adapted, however, for dairy cows, for although the milk they give is very rich, they yield but a small quantity of it, and have a tendency to get dry soon, which are marked defects in a milch cow that no other good qualities will overbalance and make amends for. The North Devon must be considered as holding a lower place than it once occupied in general estimation, being decidedly less profitable for dairy purposes than some other breeds.

The Suffolk Dun.—The Suffolk dun has long been a favourite in its own county in the districts most celebrated for dairy produce, as the cows yield a large quantity of milk. They take their name from the fact that dun used to be the prevailing colour, but it has changed a good deal latterly. They are a polled breed, of ungainly form and shape, and as stock do not enjoy a good reputation with the grazier as fattening animals.

There are a few other breeds, but those we have named are the chief races that are found in the United Kingdom, the crosses being numerous and varied between the different kinds, many excellent cross-bred animals being obtainable that will be found to suit the purpose of the dairy farmer and stock keeper better than the original breeds.

BREEDING.

It will always be found the best plan to breed one's own stock, for with care and judgment the stock keeper may obtain just those points he wishes to have, when he would often find it difficult to obtain animals to suit him exactly in every respect.

Although the dairy farmer may not want to fatten stock for the butcher, yet, as he keeps cows which drop calves with a certain amount of regularity, he may sell off his calves as store stock to others, and bring himself a considerable revenue thereby.

The best milking cows, as we have pointed out, are ordinarily not of a kind to answer the purpose of the grazier, yet some excellent cross-bred animals may be obtained by crossing either an Alderney or Ayrshire cow with a good shorthorn bull.

The issue will be a first-rate calf, if the bull is a pure-bred one, the defective qualities of the dam being made up for by the good ones conferred by the sire ; and they will turn out well as milch cows when they reach maturity, or prove to be good fattening stock, as the case may bo. Good paying animals may thus be obtained that will be found to answer the purpose of the breeder excellently well, and although perhaps the best milker may be the ugliest cow in the herd, yet her progeny, the result of a cross with a first-rate shorthorn bull, will often be a very handsome animal, possessing most of the best points of the latter. It is somewhat remarkable, also, that the smaller cows will often bring a calf when so crossed that will at maturity rival in size the animals of some of our largest breeds.

The dairy farmer may, therefore, retain the heifers so produced and add to his herd as occasion may require, or dispose of them as young stores, for which he will be enabled to realise a good price at all times, provided they are the well-bred animals indicated.

Rearing of Calves.

In some parts of the country where the calves are reared they are allowed to suck their mothers for a considerable time ; and as this interferes with the produce of the dairy, where economical expedients are not resorted to in the process of rearing calves, they are often got rid of very shortly after they are born. This is done to a great extent in the Vale of Aylesbury, upon those farms where dairy produce is made the first consideration. But where butter making is carried on there need be no falling off in the weekly supply, as calves can be brought up well enough upon the skimmed milk if boiled linseed is added to it.

Care must be taken that the milk is not sour, and it must be heated to the natural heat of milk as it comes from the cow, or, if made hotter in the course of preparation, it must be allowed to stand till it gets the proper degree of coolness, which an experienced person can tell at once by merely dipping the finger in it. There is a good deal of nicety in this, and a careful man must be employed, sufficient care being exercised that the contents of no stray milk-pan which has stood too long before skimming be given, which, although immaterial, perhaps, as far as the butter is concerned, will make all the difference to the calf, as it perhaps may cause it to *scour*.

The young animal will require to be taught to drink from the pail, and this it will soon learn to do, by the man who feeds inserting his fingers covered with milk into the calf's mouth and making a little to flow into it. The calf at first will not understand the meaning of the business, but will soon get an inkling of it, hunger teaching it pretty soon to learn to drink. But care must be taken when the calf's head is in the pail not to allow its nostrils to be covered, or, of course, it will withdraw it suddenly for the purpose of breathing.

Hay tea is also a good article to eke out the supply of milk, and meal of some kind may be mixed in it.

The calves should be kept in a dry shed, where the floor inclines to a drain, and should be kept clean and dry. An empty shed, without any divisions or fittings, is all that is required, as the divisions to separate the calves are best made with hurdles fastened down with a stake to the floor. A little sweet fresh hay should be twisted in the hurdles, and the calves will soon learn to nibble some of it; and, as they grow apace, they should be tried with a few sliced carrots, after which they will soon learn to eat and do very well. After being accustomed to be fed from the pail they will quickly get to run to it whenever they see it, but two calves should not be allowed to drink together, as one is sure to get more than its proper share, and they cannot be fed with exactitude, which is highly necessary in the case of all young stock. A careless man desirous of getting his job over as quickly as possible will allow this to be done; but such men ought never to be allowed to have the care of animals, which should only be entrusted to painstaking and considerate persons.

By this method of management the calves which drop in the early part of the year, if kept in warm sheds and carefully looked after, as the spring advances, may be turned out for an hour or two in a sunny enclosure or orchard, and be allowed to pick a little grass, taking care not to let them have too much so as to cause disorder; and, by the time the weather is fit for them to be turned out altogether, they will be comparatively strong animals, and have the whole summer and autumn before them to run upon the pastures, or pick up their living in any odd places where there is sufficient for such young animals, supplementing their feeding with a little concentrated or other food as occasion may demand. At the end of the autumn there will be a fine animal or two, which will have cost but very little to their owner beyond the care that has been exercised in bringing them up, and the value of the skimmed milk, which otherwise would probably have been consumed by the pigs.

As the full details of this kind of management really belongs more to the subject of dairy farming, and the management of cows in connection with that branch of rural economy, we will pass on to the feeding of cows, which is a very important department, that requires close looking after.

The Feeding of Cows.

Regularity in the hours of feeding cows is a point of the utmost importance, and it is a well-established fact that cows which have their food given to them with punctuality give more milk than those which are fed upon better food, but where punctuality is not observed in the hours that it is given to them ; irregularity in this respect is invariably marked by a diminution in the yield of milk.

While the cow is giving her full supply of milk her food should not be stinted in any way, and she should have a liberal allowance of everything that is necessary for her. Even when giving cows a full supply of rich food there are many expedients which can be safely resorted to for economising expense, and do the cow good at the same time. One of these is the giving of chaffed straw, especially when a large quantity of roots, as pulped mangolds or turnips, is given. The stomach of the ox is a very capacious one, and its digestive powers are so constructed by nature as to be capable of extracting a considerable amount of aliment from bulky food not rich in alimentary properties. The chaff also corrects the watery tendency of the roots, and in the early stage, when freshly drawn from the ground, corrects those acrid juices which they frequently contain. By giving, also, some cheap, bulky food, it enables the farmer to give the animals some richer food—as meal of various kinds—which is not only of advantage to the animals themselves, but greatly improves the quality of the manure they make, and a large variety of food can be given, a change being often very desirable.

Oooked linseed, bean meal, and distillers' grains, are excellent auxiliaries in the diet of the cow, but if fed too exclusively upon these without a proper admixture of chaff or hay, in time they pall upon the appetite, and, where grains have been given largely and continuously, with a view of stimulating the flow of milk, the cows have become "grain-sick." A judicious mixture of food is therefore to be recommended, as turnips, green clover, rye grass, clover, hay, oilcake, &c. ; and upon those calcareous formations which grow lucerne and sainfoin, these are excellent food for cows. The straw-chaff as an economical agent, or hay, in which a proper quantity of salt is given, also prevents accidents which sometimes occur when green food is largely used, and guards against swelling, or the disease known as hoove, or blasting, and wet succulent food should never be given without a proper proportion of dry food.

Sudden changes should never be allowed, as that of from dry food to green, or green to dry, for, after having been kept long upon the latter during a long winter, when roots even may have been getting scarce, the cows are apt to eat too greedily of the green, succulent food, and so expose themselves to injury. Half the accidents that take place in this way could be averted by the proper use of care and caution in this respect.

Roots steamed with hay and chaff are given with advantage, and at the end of summer, when the grass becomes scarce, a good substitute is to be found by steaming young turnips and turnip leaves with hay.

Good rich grass and prime meadow hay are certainly the natural food of cows, and where plenty of both of these is to be had, undoubtedly the dairy produce is of finer quality than can be got from almost any other description of food, but these are not to be had in unlimited quantities everywhere, and the larger

variety of food now at command gives the opportunity of a much
larger stock being kept than otherwise would be the case, for hay
by itself is decidedly expensive.   Cows relish a change of food,
and by feeding them on roots steamed with chaff and oat-straw,
milk has been produced, excluding other expenses, at a cost of
4d. per gallon, this being the result of careful experiment upon
the house-feeding or soiling system.

ATTENTION TO STOCK.

There is no department that will pay the young farmer so well
for the occupation of his time as being a good deal amongst
his stock, which enables him to acquaint himself with the characteristics and peculiarities of each animal.

With young stock, the calves soon get to know him, and
their docility is vastly increased by a little notice and attention.
There are many ignorant men who think that the proper way
to deal with cattle is to push them about. swear at them for
any trifling thing that does not quite accord with their own
humour, and handle them roughly.   Animals are quite sensible
of the uniform kindness or otherwise of their attendants, and
if the former prevails, they acquire that docility and placid
look and deportment which distinguishes the ox under its most
favourable aspect, a quiet and satisfied condition being highly
advantageous to their well doing and proper rumination.   By
being kindly treated, they will not manifest dislike and fear upon
being handled, when, perhaps, some little accident or ailment
may cause them to need examination and treatment.

If the calves, when they are young, are led about gently in a
halter, and petted by a handful of their favourite food being
given to them, they will be accustomed to this sort of training,
and their general management will be much easier ever afterwards.   A bad tempered man should never be employed about a
stock yard, only those of humane dispositions, who are fond of
animals, and who take a pleasure in observing them.

CHAPTER IX.

OXEN.

Varieties—West Highland Cattle—Galloways—The Ayrshire Ox —The Shorthorn—Hereford Oxen—Longhorns—Alderney— Suffolk Dun—Housing—Feeding—Warne's Compound.

VARIETIES.

FROM what we shall give in the following notes, the reader will be able to gather the principal facts relating to each breed of the leading races of oxen, and their qualifications when they are dealt with as a pure breed. Some excellent cross-bred cattle are, however, produced when the defects of one parent are neutralised by another, as in the case of the Ayrshire, the cows being good milkers, but the oxen unfit for the purpose of the stock-keeper or grazier. By crossing, however, with a good shorthorn bull, a capital intermediate race will be produced of cows that will prove good milkers, and oxen that will turn out good feeders, some of the latter rivalling in size the largest animals that are ever reared. After the breed has been chosen that it is thought desirable to select, the accommodation should be taken into account.

In grazing, there is one important principle that should be ever borne in mind by the young stock-keeper, that animals should never be taken from a rich pasture to a poor one, for they will never thrive, but when the reverse is the case, and animals accustomed to poor herbage are put upon better food, their progress is always highly satisfactory.

West Highland Cattle.—In no instance is this more palpably the case than in that of the West Highland cattle, or Kyloes, the cows of which breed we have before described as giving rich milk, but very little of it, and as having a tendency to get dry soon, so that they are unfitted for the purpose of the dairy farmer.

For the purpose of the grazier, however, there is not a better breed to be had, where the pastures are somewhat poor. In being accustomed to the coarse herbage of the Highlands and its humid climate, when the ox is removed to the more genial climate of South Britain, and to a better herbage than he has been accustomed to feed on, he will fatten where a finer breed would scarcely be able to pick up a living.

The superior quality of beef yielded by the carcase of the West Highland ox causes him to be a great favourite with the butcher, his beef commanding the highest price with private families.

For feeding down the grass in gentlemen's parks, the Kyloe has long been a favourite in England, and his bold erect carriage, broad head, with short fine muzzle, long up-turned horns, short muscular limbs, and mellow skin covered with thick shaggy hair, causes him to be a very picturesque object in such a situation. He is somewhat slow in reaching maturity, but he is well worthy of his entertainment, which is very often of a coarse and inferior description, which he has the knack of converting into so much good marketable beef. As grazing stock in many somewhat unfavourable situations, the Kyloe will be found a most useful race.

Galloways.—Adapted for pastures of rather a better description than the former, the Galloway is a most desirable animal for the grazier and stock-keeper who sell to the butcher. They are polled cattle of a larger size than the West Highland, and on this account, as well as on that of their placid, contented dispositions, a greater number can be kept in the cattle-yard than of those which nature arms with a formidable weapon of offence in the shape of large horns. As grazing cattle, perhaps, no breed can excel them, and they have long been favourites in several of the English counties, to which some breeders in Scotland regularly consign them.

There are useful black cattle also which come from the east coast of Scotland, notably Aberdeen, Angus, and Fife, both polled and horned, that are well adapted for the purposes of the grazier.

The Ayrshire.—Although we have praised the Ayrshire *cows* as being valuable for milking purposes, very little can be said in favour of the Ayrshire ox. He comes light to the scale, takes a long time to fatten, while his beef is coarse, and inferior in quality. The qualities which make the cows useful and productive as milkers end there, for as store-stock they will be found not to be profitable, but unsatisfactory. As before mentioned, however, the stock-keeper who is also a dairy farmer, may raise a capital breed by crossing an Ayrshire cow with a shorthorn bull, but he must be of pure breed, or all the advantages which might be obtained would probably not be secured, for in all cross-bred animals there is a tendency to return to the original type, for though the animal directly bred from may not present traces of any objectionable features, they may have existed on one side of his ancestry originally, and break out again in occasional instances after a generation or two, to the disappointment of the breeder.

The Shorthorn.—The Durham, or as it is now to be called, the "Improved Shorthorn," has become the established breed of cattle all over England in those situations where it can be maintained with any chance of profit, for, being a large animal, he requires better pasture than some other kinds of cattle.

About a century ago the race was nearly confined to the banks of the Tees, but its excellent qualities have caused the improved shorthorn to be held in high estimation, not only in every part of England, but all over the world.

The principal features in the breed are its extraordinary aptitude to fatten, and the enormous weight and maturity of carcase the animals attain to while only yet at the age of calves, as it were, their symmetrical form and pleasing colours ; while few other breeds furnish beef of so good a quality as the shorthorn.

The breed is said to have been originally derived from a large, somewhat coarse race—the original Teeswater—in conjunction with a yet coarser and more ungainly animal that used to be known in the East Riding of Yorkshire as the *Holderness*, which was generally faulty in the fore-quarter, with strong shoulders,

but slow and unprofitable to feed, the meat being coarse-looking
and by no means of fine flavour.

Upon these somewhat unpromising materials a long course of
care and attention in breeding, in the first place due to the
discrimination of Messrs. Charles and Robert Colling, the
celebrated breeders, has built up the points of excellence that
make up the sum and substance of the modern shorthorn.
About seven thousand head are registered every alternate year in
the *Herd Book.*

At the sale of the shorthorn herd at Tortworth, the seat of
the late Earl of Ducie, the high average price that was realised
for the different animals was due not merely to the fact that
there were a number whose descent could be traced directly from
Mr. Charles Colling's herd, but in addition to a special value set
upon the progeny descended from one particular animal in that
herd, the original Duchess. Duchess LIX., of the eighth generation from the original, fetched 350 guineas. Duchess LXIV., of
of the seventh generation, fetched 600 guineas; while Duchess
LXVI., of the seventh generation (hardly three years old), was
sold for 700 guineas. At Charles Colling's sale Comet fetched
1000 guineas, and the cows Countess and Lily, respectively,
400 and 410 guineas. These prices will attest, without further
comment, the thorough appreciation in which the breed is now
held, and has been held for some considerable time.

The chest of the shorthorn ox is wide, deep, and projecting,
with fine oblique shoulders, well formed into the chine, with
short forelegs, the upper arm being large and powerful; the
barrel is well ribbed up towards the loins and hips, and a straight
back, from the withers to the setting on of the tail, but short
from the hip to the chine, thus fulfilling the requirements of
good judges, that a beast should have a short back, yet with a
long frame—the hind quarter must be lengthy, but well filled in,
the carcase approaching as near perfection as possible. The head
of the ox is short but fine, broad across the eyes, but tapering
towards the nose, the nostrils being full and prominent, the ears
somewhat large and thin, with bright, but placid eyes, the head
being finished off with a curved and rather flat horn, and set on

a broad, long, and muscular neck, and being well proportioned, looks altogether a smaller animal than he really is.

The great point in favour of the shorthorn is his reaching maturity at an early age, a great number of beasts being now regularly slaughtered at two years old, and even under; and, in the case of four to five year old steers, it is not unusual for them to weigh 140, or even in some cases 150 stones of 14lb., large animals very often fetching £70 a piece from the butchers.

As grazing animals they require good rich pastures, and are, therefore, well adapted for some of the rich lowlands of the midland counties of England, and are also capital animals for soiling, or house feeding management ; but on poor, or thin soils, where the herbage is scanty, they will not thrive so well as some of the more thrifty races of cattle.

Hereford.—The Herefords are a breed also well adapted for fertile soils and grazing in rich pastures ; and, as mentioned previously, the breed is said by some to equal the shorthorn ; but taking into account the early maturity at which the latter arrives, the palm must be adjudged to it for being the most profitable description of stock that can be fed upon rich meadows.

Longhorns.—Much the same as the above must be said of the longhorns, which, although a capital breed, is giving way before the superior excellencies of the shorthorn, even in those counties where the most pains have been taken to keep the breed as perfect as possible.

Alderney.—The Alderney ox is quite unfitted for the purposes of the stock-keeper or grazier. After a vast amount of food and pains has been bestowed upon him, he looks but an unpromising animal after all, it being a work of considerable difficulty to fatten him. He wears an unfavourable look, too, in the eyes of a purchaser, though after being slaughtered and cut up, his carcase generally turns out much better for the butcher than it was thought for, by the appearance the animal presented before being killed.

Suffolk Dun.—The Suffolk dun is another race that has been displaced by the shorthorn, so far as regards oxen intended for the fattening stall, the cows alone being useful for dairying purposes.

HOUSING.

It is now commonly received as an established fact that, an animal warmly and comfortably housed, needs a smaller quantity of food than one exposed to a lower temperature, a certain amount of sustenance first going to keep up the natural heat of the body in all warm-blooded animals before fat is commenced to be made, and the question has lately been a good deal debated which is best for them, yards, stalls, or boxes.

Stalls are liked by many because they occupy the least space, and require but a small amount of litter, the drawbacks to the plan being that the animals require a good deal of attendance, and, perhaps, are likely to be too warm rather than not warm enough, while they do not get sufficient exercise upon the stall-feeding system.

Feeding in yards offers a good opportunity for making a considerable amount of manure, and the attendance needed by the animals is but trifling, the drawback to this method being that both cattle and manure are likely to suffer a good deal from the weather, the latter especially, where the rainwater pours down upon it from unspouted sheds, which are commonly seen in many cattle yards, so that a principal portion of the fertilising element of the manure is washed away by the rain.

The third system, the box system, is considered to unite the advantages of both these plans, as the cattle can get some exercise, they require a smaller amount of attendance than when tied up in stalls, they are protected from the cold, and feed undisturbed and in quiet, which is a great point in favour of all fattening animals.

The manure is also screened from the weather, and retaining the urine, the whole is trampled compactly down by the hoofs of the beasts, and evaporation of its volatile parts is prevented. The boxes are arranged in a roofed building, with passage room between the rows of animals, so that each may easily be got at, each box being about ten feet square, in which the beasts are not tied. To hold the accumulated manure the ground area of the box is excavated two or three feet below the ground level, and the litter is allowed to accumulate, the animal treading it into a compact mass, its height gradually increasing according to the quantity of litter supplied, which, with a fair and moderate amount, will be about at the rate of nine inches per month. Some excellent manure is thus made, which is extremely valuable to the arable farmer in his usual round of cultivation.

When cattle are kept in yards they should not be too crowded, and they should be assorted, and kept distinct as much as possible as regards age and size. Those which have been reared together will stand closer putting up than those which have been bought in promiscuously, and should one ill-tempered beast be acquired that annoys, or gores his fellows, it should be removed, for he might possibly be the means of retarding the satisfactory progress of all the rest. Half a dozen is the average number that are put up together, but this depends upon the size of the yard, and the shed accommodation, as well as the temper and disposition of the various animals, and hence more Galloways, which are a polled breed, and destitute of organs of offence, can be put up together better than a breed like their cousins, the West Highland, or Kyloes.

FEEDING.

A system of feeding used to be carried on under the old method of keeping stock, of giving as many sliced turnips to the cattle as they could eat, the troughs being carefully cleaned out, and duly replenished the first thing in the morning and at noon, plenty of good oat straw being placed in their racks daily.

Very good fat cattle have been produced upon this plan, but, as pointed out before, the capacity of the ox for stowing away bulky food is very large, and his paunch must be quite filled with something before he goes to rest and proceeds with rumination. If this capacious paunch is filled with costly food, his powers of assimilation do not keep pace with his eating capacity, and so a large amount of waste is the consequence, and when over fed with

M

rich food, the system is too much taxed, and disorder is occasioned, such as diarrhœa, if not some more serious disease. When cattle are first put upon turnips in autumn excessive purging often follows, which is worse in the case of cattle in poor condition.

This is to be obviated by the freer use of chaffed straw, which prevents the hurtful purging to which many animals are subject; but as they cannot be got to eat a sufficient quantity in its dry state, small quantities of linseed, or meal of various kinds should be mixed with it, together with some salt, and then steamed in a close vessel, or mixed with boiling water. This process makes the straw palatable, and a sufficient quantity will be readily eaten by the animals to fill their stomachs.

By resorting to this method it is calculated that one ton of roots will go as far as two tons given in the old manner, and that a bullock restricted to the use of from 80lb. to 100lb. of turnips daily, with straw chaffed, will do as well as when allowed to eat as much turnips as he liked, which would sometimes be a couple of cwt., some feeders stating that they are enabled to do as much upon 70lb. of turnips daily with 6lb. of linseed or other meal as formerly, when the supply of sliced turnips was not restricted.

Warne's Compound.—Mr. Warne, of Trimmingham, Norfolk, the originator of the box-feeding system, introduced to the notice of agriculturists some years ago what he termed a "bullock compound," for which he gave the following recipe: "Let a quantity of linseed be reduced to fine meal. Put 150lbs. of water into an iron boiler, and as soon as it boils (not before) stir in 21lbs of linseed meal; continue to stir it for about five minutes, then let 63lb. of barley-meal be sprinkled by the hand of one person upon the boiling mucilage, while another rapidly stirs and crams it in. After the whole has been carefully incorporated, which will not occupy more than five or ten minutes, cover it down and withdraw the fire. The mass will continue to simmer from the heat of the cauldron until the meal has absorbed the mucilage. On this compound being removed into tubs it must be rammed down to exclude the air, and to prevent it from turning rancid. The compound will keep a long time if properly prepared. The consistency ought to be like clay when made into bricks."

We may remark here, though Mr. Warne laid no stress upon the circumstance, that in the sprinkling of the barley-meal gradually by the hand consists a great point in the cooking. Let anyone try the familiar household dish, porridge, that has been made in this way, by the oatmeal being stirred gradually into the boiling water, and let it be thoroughly well cooked for half an hour. Persons and children who refused to eat porridge made in the ordinary manner relish the latter extremely, and declare it to be a highly palatable dish.

But to resume. Mr. Warne afterwards published an account of some improved methods which he followed in making prepared food for his cattle, in which he says: "I commenced winter-feeding this year upon white turnips, grown after flax, the tops of which being very luxuriant, are cut with pea straw into chaff, compounded with linseed meal, and given to my bullocks according to the following plan: Upon every six pails of boiling water one of finely crushed linseed meal is sprinkled by the hand of one person, while another rapidly stirs it round. In five minutes, the mucilage being formed, a half-hogshead is placed close to the boiler, and a bushel of the cut turnip tops and straw put in. Two or three handcupfuls of the mucilage are then poured upon it, and stirred in with a common muck fork. Another bushel of the turnip tops, chaff, &c., is next added, and two or three cups of the jelly as before; all of which is then expeditiously stirred and worked together with the fork and rammer. It is afterwards pressed down as firmly as the nature of the mixture will allow with the latter instrument, which completes the first layer. Another bushel of the pea straw, chaff, &c., is thrown into the tub, the mucilage poured upon it as before, and so on till the boiler is emptied. The contents of the tub are lastly smoothed over with a trowel, covered down, and in two or three hours the straw, having absorbed the mucilage, will also, with the turnip tops, have become partially cooked. The compound is then usually given to the cattle, but sometimes is allowed to remain till cold. The bullocks, however, prefer it warm ; but, whether hot or cold, they devour it with avidity."

Later on still he yet describes another variation in his method, and says : "I am now using a preparation of barley-straw with that of peas, according to the following plan : To nine or ten pails of water a bushel of swede turnips, sliced very small, is added. After having boiled a few minutes about two pecks of linseed meal are actively stirred in ; the mucilage is formed in about five minutes. A hogshead is then placed by the boiler, and one or two skips of chaff thrown in. Three or four hand-cupfuls of jelly and turnips are next poured upon it, which being mixed together with a three pronged fork, are firmly pressed down with a small rammer 3ft. long and 5in. square at the bottom, with a cross handle at the top. The first layer completed, a small quantity of the chaff, &c., is put into the tub as before, till the boiler is emptied, The mass being covered down, in a short time is ready for use."

Various modifications have been made upon the system first propounded by Mr. Warne, different ingredients being used in varying proportions in conjunction with one inch chaff, as linseed, oil-cake meal, bean meal, bruised barley or oats, &c., and some, instead of cooking the ingredients, wet them well with cold water, it being necessary that chaff when given with crushed grain or meal should be wetted.

In feeding, no animal should ever be allowed to go back, but be made to progress steadily. The force of this necessity is clearly apparent when it is considered that after a certain amount of condition has been attained, if, from bad management, the animals go back, so much food and the cost of the labour upon their attendance is actually thrown away, and, in addition, it takes a great deal more food, and often a considerable time to enable them to make up the lost ground.

In the case of young animals, neglect in regular and ample feeding is still more disastrous at times, for they acquire a stunted habit of body, from which, perhaps, they never after-wards recover.

There are, however, various ways of feeding cattle effectively, and yet economically, which each should endeavour to find out for himself, as being best fitted for his own peculiar circumstances, the food at command, and what crops suit him best to grow.

Flies and insects in warm showery weather are very troublesome to cattle, and it will always be found a good plan to keep the animals under cover during the hottest period of the day during summer, and turn them out when it is cooler, and, if this cannot very conveniently be done in the case of rough store stock that may be at a distance from home, the milch cows might, at all events, be thus taken care of, to the manifest improvement of their comfort and well-being.

CHAPTER X.

THE DISEASES OF CATTLE.

*The Diseases of Oxen—Abortion, Slipping Calf, or Warping—
Blain—Blasting, Hoove, Hoven, or Meteorization—Bronchitis
—Catarrh—Choking—Cow-Pox—Loss of Cud—Diarrhœa,
Scouring, the Scant—Drop in Cows—The Epidemic—Hœma-
turia—Inflammation of the Kidneys—Disease of the Liver—
Inflammation of the Liver—Loo, Low, Foul in the Foot—
Moor-Ill, or Wood-Evil—Paralysis or Palsy—Pleuro-Pneu-
monia— Quarter-ill— Redwater—Rheumatism—Distension of
the Rumen — Inflammation of the Rumen—Skin Diseases:
Mange, Lice—Thrush—Diseases of the Udder—Inversion of
the Uterus and Vagina.*

A GREAT many of the diseases to which oxen are subject are to
be averted by careful attention to housing and feeding.   In some
of what are called the best dairying counties, as Gloucester, the
cows are allowed to be exposed a great deal too much during
inclement weather in the open fields, there being throughout the
county generally a great deficiency of house accommodation. By
being exposed too much, cattle get rheumatism, catarrh, and
bronchitis, and in damp low-lying districts cows are exposed to
the chance of getting diseased udders.

There are a good many casualties that happen to cattle in
connection with the processes of feeding, and derangement of the
digestive organs.

Abortion, Slipping Calf, or Warping.—Abortion is commonly
supposed to mostly arise from the occasion of accident, but it
takes the form of a disease as well, for cows that have once
aborted are very likely to do so again, and it has been known also
to prevail amongst cows more at particular seasons, resembling
somewhat the course of an epidemic.

Cows that do not breed early are more likely to slip their calf
than young heifers, and it usually takes place between the ninth

and fifteenth week, but may occur at any period. It may be
brought on by blows, strains, or jumping other cows, or being
hunted about, or by any kind of fright, as well as by disturbance
of the digestive organs.

If it takes place during an early period of the cow's pregnancy,
the consequences are not often very serious, but when at a late
period critical consequences may ensue. Opening medicine is
usually given first, followed by sedatives, an ounce each of
laudanum and spirit of nitrous ether following a dose of salts.
When inflammation is expected, hot fomentations should be
applied to the loins.

If symptoms of likely abortion make their appearance, prompt
treatment should be resorted to, and the animal removed to a
quiet place and bled, and have administered to it an ounce and a
half each of tincture of opium and spirit of nitrous ether, but
no aperient medicine, and if a cow has slipped her calf before at
a particular time, this should be carefully noted, and she should
be bled just previous to the corresponding period when this took
place.

Blain.—Blain consists of an inflammation of the membrane
lining of the mouth and tongue, also called *Gloss Anthrax*, the
mouth swelling and becoming covered with blisters, so that in
bad cases the animal is unable to eat. Early treatment is
imperatively called for, and it may be necessary to lance the
tongue and take some blood from the roof of the mouth, while
a dose of salts should be carefully administered. Nitre and
tartarised antimony in small quantities should be given in gruel,
and the mouth washed with a healing lotion formed of the
following :

Powdered alum...	2dr.
Honey	1oz.
Sulphate of zinc	1so.

in 1lb. of warm water—*i.e,*, about one pint.

Blasting, Hoove, Hoven, or Meteorization.—This, which is by no
means an uncommon disease where cattle are changed so quickly
from dry food to rich succulent green food without any intermediate preparation, consists of distension of the rumen, due
to the gas given off by the food in its fermentation, the stomach

being often distended to an enormous size, and suffocation soon follows unless relief is given.

The common practice is to pass a hollow flexible probang into the stomach so as to allow the gas to escape through it. A draught either before or after the application of the pro-bang should be given to condense the gases through means of chemical reagents. The draught may consist of 1oz. of harts-horn and 3drms. of powdered ginger in a pint of water. If medicine is not readily obtainable, lime water is a useful remedy to apply, or a couple of drachms of chloride of lime dissolved in a quart of water. A purgative should be given afterwards to restore the tone of the digestive organs. Sometimes an incision has to be made in the flank by the trochar, in which a quill or stick of elder is inserted, through which the gas escapes.

Bronchitis.— When catarrh has been neglected sometimes bronchitis may set in, when greater soreness is apparent in coughing. A seton should be inserted in the brisket, and bleeding resorted to, and aperient and febrifuge medicine given.

Catarrh.—Exposure to the weather brings on, first, catarrh, afterwards bronchitis, and frequently rheumatism, while expo-sure to hoar frosts often brings on a sudden and fatal disease to which young yearly heifers are especially liable, called "Quarter-ill," "Blood-striking," or "Black-quarter."

Catarrh amongst cattle mostly prevails in the spring of the year, when the seasons are wet and cold, and particularly when east winds prevail—young, unseasoned animals being oftener affected than mature beasts.

Good beds and dry housing, together with a few warm bran mashes, will generally effect a cure if taken in time; but should the attack be a sharp one, moderate bleeding is recommended, and a dose of Epsom salts. In severe cases it may be desirable to rub a liniment into the throat, composed of :

Olive oil	6oz.
Powdered cantharides	1oz.
Oil of turpentine	2oz.

Sometimes catarrh will turn out to be of an epidemic nature, and upon these occasions there is a greater tendency to debility, vegetable tonics—as gentian and ginger—can be given with advantage.

Choking.—Choking is a very common occurrence with animals that are fed upon cut roots when not sufficient care is used with them, a piece of turnip, mangold, or potato getting firmly fixed in the œsophagus, which, pressing upon the windpipe, interferes with respiration. The probang, with a knob at the end, which should be first well oiled, in most cases needs to be used. It is also a good plan to give the affected beast a little oil as well, from the horn. The probang should be firmly but gently pressed against the obstructing object, the animal's head being alternately depressed and raised the while. If it cannot be made to pass at once, undue haste should not be used, because, at a second attempt, it may be made to pass easily, and care must be taken not to lacerate the lining membrane of the œsophagus and its muscles, or serious consequences afterwards may ensue.

By giving pulped roots and steamed food, the dangers arising from choking are a good deal lessened. Soft food should be given for a few days after an operation of this kind has been performed, so as to allow the overtaxed muscles, which are often a good deal strained, to resume their wonted condition.

Sometimes the offending body will become firmly impacted in the roof of the mouth, in which case it will need to be withdrawn by the hand upwards.

Cow-pox.—This is now rarely met with, but the pustules which form on the udder and teats will give way to the use of astringents, such as a little powdered chalk and alum, while a cooling aperient should be given. It is infectious, and can be communicated from one cow to another by the hands of the milker.

Cud, Loss of.—When this takes place, it is more a sign of there being something wrong with the system than the indication of a disease itself, and mild purgatives with tonics should be given to restore the tone of the animal in the absence of any other indications of disease.

Diarrhœa, Scouring, the Scant.—Diarrhœa is mostly produced by improper food, or by a too sudden change from dry to green.

The most simple form of the disorder consists of a relaxed condition of the mucous coat of the small intestines, but it will sometimes indicate a graver disorder, as disease of the liver or of the maniplus.

A change of food will often effect a cure in the first place, but if the looseness continues some gruel should be given, to which is added :

Powdered gentian root	2dr.
Opium	½dr.
Prepared chalk	2oz.

and given once or twice a day, or as may be deemed necessary.

If the liver is thought to be affected, calomel must be given, which is best associated with opium, and $\frac{1}{2}$ drachm of each administered twice a day. When there is reason to think there may be some irritating substance, it will be prudent to clear out the bowels with Epsom salts.

With young calves scouring often takes place, owing to coagulation of the milk which they suck, which causes derangement of the stomach, the whey passing on and causing purging, when the evacuations are of a white colour. A great many calves are lost from this disorder, the coagulated milk forming a large mass, which it is necessary to dissolve by the administration of medicines of alkaline properties, such as carbonate of magnesia and carbonate of soda, which should be given in doses of one or two drachms each, according to the size and age of the calf, so as to neutralise the acids that most likely exist in excess in the stomach.

Drop in Cows.—The drop is a sudden disorder, which sometimes strikes down a cow shortly after she has calved, and when to all appearance she is doing well. It is considered to be an affection of the spinal marrow at the region of the loins. The animal so struck down often lies stretched out without any signs of animation, and as the drop seldom takes place till after a cow has had several calves, the theory of the disorder is that with each the uterus becomes more dilated, and consequently the contractions are greater, and the muscular effort in expelling the fœtus, combined with these involuntary contractions, produce it.

There is often no room for treatment, and the animal dies,

but in curative cases both purgatives and stimulants must be used, which can be given in the form of the following, in oatmeal gruel :

Flour of sulphur	4oz.
Sulphate of magnesia	1lb.
Croton oil	10drps.
Spirit of nitrous ether	1oz.
Carbonate of ammonia	4dr.
Powdered ginger	4dr.

the object being to restore the loss of action.

When there is obstinate constipation the dose of croton oil may be increased, and from 8gr. to 10gr. of powdered cantharides given, as well as rubbing a blistering liniment on the spine and loins. A fourth of the above medicine should be given every six hours, omitting the croton oil until purging is brought on, and plenty of nourishing gruel should be given, as well as bran mashes, if the patient can be got to eat them.

Injudicious feeding, in our opinion, often brings on this disease, and the precaution should always be taken before a cow calves to give her a good dose of Epsom salts, for sometimes the stomach may be loaded with food difficult of digestion, or the system be in a plethoric condition, when the muscular efforts made in the course of parturition must be necessarily greater from the stomach being distended by food. The same effect is also produced from too plethoric a habit of body.

Epidemic, The.—This is rather a confused term when applied to cattle, and is identified with the *murrain* and *blaine*, and, although not a fatal disease, is yet a very serious one in its results to the farmer, as the flesh of the cattle decreases, and, the supply of milk in cows falls off. Though sometimes slight, and lasting but a short time, at other times it is dangerous, and continues for a considerable period. The first symptoms are a cold fit, with staring coat and cold extremities, which is followed by a reaction, the extremities becoming warm, and a discharge of saliva issues from the mouth. The muzzle becomes dry and hot, feverish symptoms are exhibited, and vesicles form on the tongue, parts of the mouth, and in the case of cows the teats are often thus affected. The beasts feed but slowly, sometimes from want of appetite, but very likely more often from pain in eating. The nature of the disease being that of a low fever, and tending to produce debility, it is improper to bleed largely—which is sometimes done, with the view of cooling and reducing the system—for occasionally this causes the disorder to take the form of typhus, under which the animal soon sinks.

The correct treatment consists in moderating the fever, and using astringents to the mouth and feet, which latter are often affected, and then to support the strength of the animal by tonics. Epsom salts with sulphur make a useful aperient, and gentian root, ginger, and sulphate of iron—two drachms of each—make a good tonic. If the udder is affected, bleeding in the veins and hot fomentations will be necessary. Should the liver or lungs be affected, as will sometimes be the case, proper remedial treatment will need to be resorted to.

Hæmaturia is produced by strains, sometimes occasioned by cattle riding one another, being a rupture of small vessels in the urinary passages. Medicines, as in Redwater, are administered, and if these do not have the desired effect, the following is given in linseed gruel :

Oil of juniper	2dr.
Oil of turpentine	1oz.
Tincture of opium	1oz.

Kidneys, Inflammation of the.—Nephrites, or inflammation of the kidneys, is somewhat rare, but is sometimes produced by cold and wet, by strains, or even blows. Its presence is indicated by weakness in the loins, and a discharge of dark coloured urine. Bleeding is advisable, and a purgative should be given the same as recommended for Redwater, and the loins be well stimulated by a mustard poultice, and linseed gruel given.

Liver, Disease of the.—In a state of nature, the liver of an ox is seldom diseased, but high feeding with artificial foods will sometimes produce it. Epsom salts should be first resorted to, combined with carminatives, and if these do not answer, a drachm of calomel should be given in gruel.

Liver, Inflammation of the.—An abundant supply of nutritious food, perhaps, combined with a plethoric disposition, will sometimes bring on inflammation of the liver, or hepatitis ; as will also exposure to heat, undue fatigue, &c., by which the general system is disordered.

The treatment must depend a good deal upon the condition of the affected animal, but calomel in doses of a scruple should be given in conjunction with opium, the same quantity—or even two scruples of the latter—twice a day, and the bowels should be cleared out with Epsom salts.

Low, Loo, Foul in the Foot.—This disease commences with inflammation and lameness, succeeded by soreness between the toes, the foot discharging offensive matter, in its aspect resembling somewhat the foot-rot in sheep ; successive abscesses sometimes form. The disease most frequently occurs upon damp or marshy soils, and is supposed to arise from the softened condition of the feet and the friction of the mud between the claws.

A pledget of tow, dipped in tar, over which some powdered sulphate of copper may be spread, should be inserted between the claws and renewed when necessary, which may probably be in forty-eight hours.

Moor-ill or Wood-evil.—This is a singular disorder that some-times visits cows that graze in the neighbourhood of woods and commons, and is manifested by a depraved appetite, the animal taking up stones, bones, pieces of iron, or any stray substances that may come across her way, going about with her belly tucked up and losing flesh daily. The bowels are invariably obstinately consti-pated, and she appears to move with pain, which is manifested by groans, the secretion of milk almost ceases, as does also rumi-nation, for very little food is eaten. Opening medicine should be given, succeeded by febrifuge, and alterative medicines, and if there are symptoms of inflammation of the lungs, bleeding may be necessary, but not without ; and every pains should be taken to get the digestive powers into order, which are altogether thrown out of gear. A seton may be inserted in the dewlap, and occa-sionally the sides may be blistered as well.

Paralysis, or Palsy.—This is a disorder frequently met with amongst cattle, being familiarly called "Tail-slip" as well, the symptoms being an apparent inability to raise the tail in making the usual evacuations, by which the hinder parts of the animal become filthy. It generally proceeds from rheumatism in the first place, and is more common with cows and young beasts that are poorly fed, seldom visiting stall-fed beasts. It is sometimes associated with inflammation of the membranes of the heart and chest.

A good deal of ignorant superstition used to prevail with regard to this disease, one of which was that it was owing to the presence of a *worm in the tail.*

An affected animal should first be comfortably housed, then bled, the loins blistered, and a seton put in the dewlap. Purgative medicine is given, combined with carminatives, and the animal fed upon good nutritious food.

Pleuro-Pneumonia.—A great deal of difference of opinion has prevailed as to the true nature of this scourge, which is well recognised under the scientific name it bears, and which comes under the notice and control of the veterinary department of the Privy Council, but the term denotes inflammation of the substance of the lungs and also of the membrane which covers them.

Its infectious character is very manifest, which it is thought by some is due to an animal poison floating in the air, which in most cases proceed from the respiratory surfaces of diseased animals.

In the early and slight stages of the disease there is a short and slight cough, and the coat of the animal is somewhat "staring." A singular circumstance in connection with the disease is that the most valuable kinds of stock, as the shorthorn, appear to be amongst the readiest sufferers.

The rule adopted is not to wait for treatment, but to slaughter the affected animals at once, even on suspicion, if the herd is large, but when it unmistakably has made its appearance, the Government inspector will solve the difficulty at once, and spare all further trouble on that head.

In cases of suspicion, the animal or animals should be separated at once, and internal remedies given in the shape of Epsom

salts and linseed oil, to regulate the bowels, and then administer
a sedative composed of the following:

Nitrate of potash	?dr.
Powdered white hellebore	1dr.
Tartarised antimony	1dr.

made up into a powder and mixed with gruel, and given at
morning and evening of the first day of treatment, and once a
day afterwards for four or five days.

Quarter-ill.—This sudden and often fatal disease is also known
by the names of "Black-quarter" and "Blood-striking," the effects
of opposite causes being included under these names. As we have
before mentioned, young cattle, and especially yearly heifers,
acquire the disorder from lying on a cold damp soil, especially
when there has been a hoar frost, which perhaps for the first
time comes as a shock to the system of the young animal. It is
so rapid in its effects that an animal well on one evening may be
found dead the following morning; but generally the affected
beast is found with one quarter very much swollen, and lame.
The animal should be housed at once, and a stimulant administered, consisting of a drachm of camphor and a couple of ounces
of spirits of nitrous ether, given in gruel. Bleeding is sometimes
practised, but this should not be done if the pulse is feeble.

The same apparent form of disease occurs at quite a different
time of the year, and sometimes visits two-year-old cattle when
they are put upon rich pasture, and a sudden change made from
poor winter food. The animals should be bled, the bowels well
opened, and the affected parts fomented with hot water, an
access of blood causing the disease in the latter instances.

Redwater.—This is also a disease of the liver, principally
caused by the digestive organs being out of order, the urine
being charged with biliary deposits. The bowels should be first
well opened with a draught composed of the following:

Calomel	1sc.
Sulphate of magnesia	12oz.
Sulphor	4oz.
Powdered ginger	3dr.
Carbonate of ammonia	4dr.

given in gruel.

A quarter of the above may be given every six hours, leaving out the calomel until the bowels are relaxed, when mild stimulants and diuretics should be given, made after the following recipe :

Sulphate of potash	2dr.
Gentian root (powdered)	1dr.
Ginger ditto	1dr.
Spirit of nitrous ether	1oz.

Rheumatism.—This affection, known also locally as "Joint felon" and "Chine felon," is generally contracted from exposure to the weather, and may be acute or sub-acute, affecting either the fibrous tissues, or extending to the muscles and sinews, sometimes attacking the serous membrane lining the chest and affecting the heart. There is often a good deal of fever, and the usual symptoms of stiffness and pain in moving, the joints being generally affected when the disease is sub-acute.

Bleeding is generally resorted to, and a purgative given, in which is mixed 1oz. of the spirit of nitrous ether. The ether may be repeated twice a day, together with a drachm of colchicum and another of tartarised antimony. It gives ease to foment the suffering parts with warm water, afterwards rubbing into them the stimulating liniment mentioned under catarrh.

Rumen, Distension of the, by Solids.—Distension of the rumen, though not so sudden or alarming in its appearance as hoove, is yet a more serious disorder, being principally met with in the case of stall-fed beasts, and arises from accumulated food getting so hard and dry that it can no longer be returned to the mouth for second mastication. It is sometimes difficult to find out the exact state of the case, and this is at times tested by moving the trochar to and fro. In urgent cases it is sometimes necessary, in order to save life, to make an opening into the flank through the rumen, large enough to allow of the contents being removed by the hand, but when the case is not serious, a drench composed of purgative and carminative medicines is given, assisted by injections. Sometimes the stomach-pump is resorted to, and liquid injected to excite vomiting. When there is anything like an operation necessary, the services of a veterinary surgeon need to be called into requisition.

Rumen, Inflammation of the.—This is comparatively a rare disease, and mostly arises from the animals partaking of some

poisonous plant in the pastures which, as a rule, their fine sense
of smell enables them to detect and refuse. The disease, called
ergot, which is most conspicous in rye, but infests other grain and
grasses, excites abortion in cows, and sometimes causes death.
In these cases the stomach-pump must be called into requisition,
and oily purgatives and injections given. The clippings of yew
trees, which are poisonous, are sometimes eaten by cattle.

Skin Diseases—Mange—Lice.—Mange may be acquired by
contagion, or produced by poverty of condition, and is due to the
presence of an insect termed the *acarus*, which breeds under the
skin and causes intolerable itching. It may be cured by rubbing
into the skin, with a good deal of friction, an ointment composed of

Sulphur vivum	4oz.
Linseed oil	8oz.
Oil of tarpentine	2oz.

Lice are generally a sign of poor living, infesting cattle that
have been reduced by hard fare, and may be destroyed by the
above ointment or by using tobacco water.

Thrush.—The thrush, or *apthœ*, consists of small pustules
which break, become sores, and heal again in about ten days'
time, in the membrane lining the mouth. A dose of Epsom
salts should be given, and a weak solution of alum and water
applied to the mouth.

Udder, Diseases of the.—Where cows have been exposed to
cold and wet, disease of the udder is sometimes contracted, the
part swelling, and becoming hot and hard, while the secretion of
milk is interrupted. Hot fomentations should be applied and
opening medicine given, but, if shivering comes on, a stimulant
should be administered in the form of an ounce of ground ginger
in some gruel or warm ale, together with 2oz. of spirit of nitrous
ether.

After the udder has been well fomented, ointment, composed
of the following, should be rubbed into the part:

Mercurial ointment	2dr.
Powdered camphor	1oz.
Lard	½lb.

well mixed up. If disease is allowed to proceed unchecked, it
may end in the loss of one or two quarters of the udder.

N

Uterus and Vagina, Inversion of the.—Inversion of the uterus generally occurs after parturition, while the latter may take place before, in both cases it being necessary to wash the parts carefully, and restore them to their proper place as quickly as possible, keeping the hinder parts of the animal higher than the fore ones.

When calving is prevented by an unusual presentation, the calf should, if possible, be restored to its natural position, which is with its head resting on the fore legs, as these ought to be presented first.

In cases of what are termed unnatural presentation, sometimes the fœtus literally has to be removed piecemeal to save the life of the cow, and when the buttocks are presented first, care should be taken to cause the hind feet to escape in the first place when it can be managed.

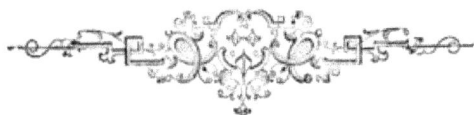

CHAPTER XI.

ASSES, MULES, AND GOATS.

The Domestic Ass—Varieties—Advice in Buying—Feeding—Breeding—The Mule—The Hinny—Uses—The Goat—Advice in Buying—Uses.

THE DOMESTIC ASS.

IT is often a matter of surprise, that the domestic ass in England has been made so little use of. He can be purchased for about one-twentieth of the price of a horse, and can be kept at a most trifling expense. It has been remarked by a writer, "we are at a loss to know how to estimate the expense of keeping the ass; for so long as there is a hedgerow overgrown with briars and thistles, so long as there is waste land furnishing a few tufts of rank and bitter grass, the rejected of other cattle, but to our poor friend welcome, so long will the ass stick to his work, thrive and cost you nothing.

"But we recommend that the ass should cost something. He ought to receive better treatment than this; he will repay a daily allowance of hay, or permission to graze in better pasture; for consider what he is, a powerful and patient drudge, rarely the subject of disease, long-lived, and fit for his work almost to the last.

"In draught, two good asses will knock up any horse. In the plough, especially in light soils, they are sufficiently effective; and both male and female produce an intermediate creature by intercourse with the horse or mare respectively, by many degrees more valuable for this purpose than either parent. Many objections have been urged against the ass. He has been called obstinate, slow, and mischievous. Those accusations may be true, but we should blush to bring them forward, for to our mismanagement and cruelty they are alone to be attributed.

Three generations would produce him a different creature.   Give him proper care, afford him only half as much attention as would freely be bestowed upon the most worthless horse ; and there is little doubt but that in a very few years the animal would rival, in every respect save fierceness, those coursers of the desert, described by the accurate Morier.   His bad qualities are the result of the neglect and cruelty of his treatment.   Kindness and attention will remove them.   The writer of this memoir can state from experience that the ass is as capable of as much enduring affection and docility as the horse."

It may well be a matter of surprise that asses are so little employed by farmers.   No farm ought to be without its donkey and little cart to perform the many odd jobs which they are so capable of doing.

In carrying small loads, collecting manure, weeds, in moving hurdles, or in the thousand and one jobs that are to be done about a farm, they will be found most useful.

Varieties.—The ass known in England is of an inferior description, and has been very much neglected, but his capability of improvement has been proved over and over again, some ladies having driven them in low carriages very successfully, at a good rate of speed, when they have manifested none of the bad qualities with which they are generally accredited.   Doubtless they might be greatly improved by crossing.   The wild ass of Persia and of Africa is an animal of great speed and power, while there is a race of Arabian origin chiefly used for the saddle, and those reared in the Island of Gozo, in the Mediterranean, have reached the height of fourteen hands ; a few of these might with advantage be brought to this country.

The Spaniards possess a fine breed of asses, a good deal of attention having been paid to them, a royal stud of stallion asses having been maintained at Reynosa, in the Asturias.

Advice in Buying.—It is very seldom that much pains is taken when a purchase is made of a donkey, yet there are very prominent points in the animal, which an intending purchaser ought to look for, and obtain if possible.   A good ass should be of fair stature, as large indeed as it is possible to find one.   The neck should be long, the nostril wide, with large full eye, and withers raised, in which latter particular asses are very faulty, back full, with large quarters. A short tail is an indication of strength and vigour, though on the score of appearance it would not be preferred perhaps.

Feeding.—As we have previously said, the ass is contented with the poorest and hardest fare, no food coming amiss to him, and he will thrive upon the very commonest kind of fare usually supplied to horses.

Breeding.—The she ass carries her young a few days over eleven months, and the young animal does not arrive at maturity till his fourth or fifth year; but as the ass will breed with the horse, a valuable creature may be produced in the form of a mule, the progeny of the male ass and the mare.

When the pairing is reversed the offspring of the horse and she ass is termed the "hinny," and in some parts of the country a mute, in Ireland the "gennatin," the hybrids being very different from each other in size and form.

THE MULE.

When well and carefully bred, the mule is a most valuable creature, being large in size and swift, taking the nature of the horse rather than that of the ass, except with respect to ear and tail. It has the quick paces of the horse, with the patience, strength, and endurance of the ass. A Spanish jackass with an English thoroughbred mare produces the best mules. In the southern provinces of France and Spain the mule is used to a very great extent as well as in the East.

They are more hardy in constitution, more muscular in proportion to their size and weight than horses, and as well as being more patient are less subject to disease than horses. They are also long-lived, being usually able to work from thirty to forty years.

Yet, although wherever they have been employed in this country regularly, their utility has been amply demonstrated, there is an unaccountable prejudice against using them.

The hinny is smaller in size than the mule, less robust, and of inferior value.

Uses.—The remains of the ass when dead are put to several useful purposes. The integuments are employed for making parchment, and the hide is of considerable value for making shoes. The substance known as shagreen, but more correctly sagri, is also made from it. The best material for drums is said to be of the parchment manufactured from the integuments of the ass.

THE GOAT.

The goat is so similar to the sheep in its main construction, the skeleton of both resembling each other closely, that a very slight difference is thought to exist between them. Goats were formerly largely kept in Wales, where the hilly districts are peculiarly suited to their nature and habits.

As an animal devoted to the production of food, it is but of small value in England, there being a great prejudice against the flesh of kid, though they are regularly consumed in southern countries; but the she goats are valuable for the milk they give, which is often much affected by invalids. A good goat, when in full milk, will yield a quart of milk three times a day, at morning, noon, and night, that is, half-past seven or eight o'clock of the evening, for goats should be milked thrice a day, in consequence of the small capacity of the udder. Goats give more milk if tethered to a certain spot than if allowed to roam about at will, and they can be grazed on almost any rough or waste ground. Like the ass, they will thrive on very poor fare, although they pay well for a more generous living.

The goat is a useful little animal *if of a good sort*, for there are some which will not give the third of the quantity of milk we have mentioned. A good goat will give milk all the year round, up to within a few weeks of parturition.

The goat breeds once a year, going to the buck in December, and bringing forth in April, carrying its young about four months, and usually bringing two, and sometimes three, kids at a time.

Advice in Buying.—The goat should be of the largest size. having hard and stiff hair, but not too abundant. Neck short, thick, and resembling that of the sheep. Head small and narrow about the muzzle, eyes large and full, and those which are hornless are generally the best milkers. Colour as dark as can be found, those of a light yellow, or pied colour, not being so good. The legs should be straight, with even and firm joints, and the ears large and full.

Goats are best at the age of from three to six years. A goat usually goes to the buck when six or nine months old, but the milk she gives at her first kidding is so trifling in quantity as to be scarcely worth speaking of. It is also better to allow a goat to suckle her first kid, as it increases her supply of milk, and she will more than repay for it the following year.

Uses.—Cheese, as well as butter, has been made from goats' milk, but both possess a peculiar taste, not welcome to all palates. The hide is dressed and used for many purposes, amongst others for gloves, and is admirably adapted to the manufacture of the finer description of shoes, being durable as well as soft and elastic.

The hair, which may be clipped annually, about the middle of May, possesses the quality of being indestructible in water, ropes made from goats hair bearing all weathers, and never rotting from moisture. It is also woven up in several kinds of textile fabrics abroad.

For further particulars of goats and their management, we refer the reader to Stephen Holmes's "Book of the Goat."*

Farming lately has appeared to be at a discount, probably to a great extent owing to a succession of bad seasons; but we are sanguine that farmers will take heart again, and that with more favourable weather, and by the adoption of the economical methods of feeding stock which we have recommended in the foregoing pages, they will see a return to more prosperous and profitable times.

---

* The Book of the Goat, containing Practical Directions for the Management of the Milch Goat in Health and Disease. Illustrated. By Stephen Holmes. Cheap Edition, 1s. London : *The Bazaar* Office.

INDEX.